GETTY

Robina Lund

GETTY

THE STATELY GNOME

with
drawings by the author
photographs by Catherine Lund

M. & J. HOBBS
in association with
MICHAEL JOSEPH

First published in Great Britain by
M. and J. Hobbs
25 Bridge Street, Walton-on-Thames, Surrey
and Michael Joseph Limited
52 Bedford Square, London WC1B 3EF
1977

ISBN 0 7181 1616 X

Set and printed in Great Britain by
Tonbridge Printers, Ltd, Peach Hall Works, Tonbridge, Kent
in Baskerville twelve on thirteen point on paper supplied by
John Dickinson & Company, and bound by
Dorstel Press, Harlow

Contents

The author wishes to thank the following: the *Daily Express*, London, for permission to reproduce the 'Millionaire' competition in chapter 10; her mother, Catherine Lund, for permission to use the photographs herein, and her mother (writing under the name of Catherine de Touchet) and the publishers, Arthur H. Stockwell Ltd. of Ilfracombe, Devon, for permission to quote from *Terse Verse*.

Illustrations

TO CATHERINE,
MY MOTHER AND MY FRIEND

'Quanti est sapere! Nunquam accedo ad te, quin abs te
abeam doctior!'

('What a great thing it is to have wisdom! I never come to
you without going away wiser!')

Terence c.190–159 B.C.

The Best Cellar

Don't mint me a million
I care not for riches.
Don't carve me a statue
They're strictly for niches.
Don't cage me a peacock
I care not for cages.
Don't write me a sonnet
That goes on for ages.
Don't mine me a diamond
I care not for flashes.
Don't win me the Test Match
I've no hearth for Ashes.
Don't tempt me with titles
Or the richest man's crumbs;
Just fill up my cellar
With Getty's oil drums.

Catherine de Touchet

Preface

In May 1965, Paul Getty was sitting in the study at Sutton Place, his chair surrounded by papers, letters and books piled high, as usual, on the floor. He was reading the latest news-clippings about himself, neatly placing each, as he finished it, on the arm of his chair. The last one was long – an account from an American paper of what purported to be his views on business, politics, marriage and money, based on an interview he had given a few weeks earlier. Eventually, he added this to the rest, took off his gold-rimmed spectacles and gazed pensively out of the window.

'There's a curious thing I've noticed,' he said at last. 'All these articles and biographies about me seem to describe some sort of computer-being, pulling the strings of countless puppet enterprises while golden coins cascade in ever-increasing heaps around it; alternatively, I'm portrayed as the solitary figure purchasing works of art in all corners of the globe by picking up the telephone and instructing some anonymous underling to purchase for me three Rembrandts, a few Canalettos and a couple of Rubens, simply because I have a few empty spaces to fill on my walls! Neither assessment makes me out a human being: I'm an object of curiosity viewed warily as not quite normal;

I'm psycho-analysed, my motives questioned, my alleged pronouncements dissected – and I come out of it all quite unlike how I feel I am!'

'When people come to interview you, most come with preconceived ideas,' I said to him. 'They expect you to be the unimaginably wealthy businessman who is also an art collector, so their questions are framed on that basis. You perpetuate that image by your answers because you tell them what you know they want to hear, which may or may not be the whole truth. They expect you to be concerned only with business and in art collecting, so they don't even ask you what other interests you have. They expect you to be miserable, because that is how you are always described, so they don't trouble to find out even whether you have a sense of humour, much less what amuses you. But, having said that – how would you like be described for posterity?'

'I would like to be remembered as a footnote in history,' he said almost wistfully, 'but as an art collector, not a money-laden businessman! I would also like people to know that I am not always "miserable"; it is simply, as you say, that I am expected to be solemn and serious. If I went around all the time with a grin on my face, people would soon start saying to each other: "What a silly fool he looks!" But, as you know, I do like to laugh and fool around sometimes.

'I also get very tired of all these comments about my "lady friends". It is very difficult for me to have a close man friend. What the public doesn't realise is that most men are envious and jealous of me, either because of my so-called luck in business – which has been nothing more than very hard work over many years – or because of my equally so-called luck with women.

Consequently, my closest men friends have almost always been very successful in their own field and happily married, which usually doesn't leave them much time to come and visit me. Since I think I have as much right as anyone else to relax with friends, many of those friends must therefore inevitably be women! It's a little unfortunate that a lot of women automatically go out of their way to try to be attractive to me – but I don't kid myself about them. They are attracted by my reputation for money and power – not by my handsome face or noble character! And the majority of very wealthy single men will tell you sadly that most, though not all, women can be divided into two types: those that you pay to stay with you and those that you pay to stay away! There are a few, of course, who are exceptional; they will stay from loyalty or go, from unfaithfulness, whether or not you pay them, thank them or complain at them! So much for my choice, if such it be, of friends!'

From this conversation evolved this book. It grew, slowly at first, from both of us remembering anecdotes, small incidents and remarks of Paul's, which one or other, or both of us, thought showed his 'other' side – that is, the one not generally known.

Soon we had so many stories that I started to write them down in note form although at that stage I had no clearly defined idea as to what we were going to do with them! In a way, collecting them was half the fun and Paul felt that, as they were about him, I should act as a sort of Boswell to his very humble version of a Johnson.

The day came, of course, when one or other of us remarked that there were almost enough notes to fill a book and, suddenly, Paul said, 'I wish you would write that book for me, Robina dear, but not while I'm alive. It reveals a great deal about me that I've realised, after all, I would prefer to keep away from other people at present. Write it for me as a memorial when I'm dead – and before the other would-be authors leap in with what I think are called "hatchet jobs", tearing to pieces any reputation I may still have left!'

After some hesitation I agreed, but subject to certain conditions. In order to retain its authenticity and accuracy as far as was possible, I insisted that nothing should go in that had not been agreed by both of us – and that included not only the anecdotes themselves but the individual facts and reported conversations – and that nothing should go in that I had not personally seen or heard and for which I could not therefore personally vouch (which will explain why I and my family are such frequent participants in his 'adventures').

This is not a biography and it is not a complete portrait of Paul, but a profile only. There are no confidential disclosures or scandalous exposures; no secret hints on how to make a million dollars; no earth-shattering pronouncements on the future of mankind.

My years of note-keeping have produced enough material to fill another book but it would be vastly different from this, showing the darker shadows of his many-faceted character. This volume, however, is about his lighter side, the laughs and the escapades with a sprinkling of annoyances to keep the balance! One had to learn tolerance in order to exist in any kind of proximity with Paul, and I hope that these stories will be read and Paul judged with a similar tolerance.

Inevitably, there are those who will ask why should I write the book? Why should Paul have asked me when he knew so many professional journalists and had so many friends? The answers he supplied himself: because I was in a unique position. I was a friend, a family friend but not one of the famous 'girl-friends'; because, although I worked for him, I was not a full-time employee but an outside 'consultant' with my own business concerns quite independent of him and, lastly, because he and I shared such a remarkable number of interests.

I grew up in a family to whom the constant acquisition of knowledge is a delight and a vital food for our enquiring minds. With, in particular, my mother, I have visited museums, art galleries, stately homes, exhibitions, theatres and auction rooms. We have travelled, read voraciously and discussed all manner of topics from geophysics to politics to graphology.

In Paul, I found the first person I had met outside my own family who shared a similar breadth of inquisitiveness and who, literally, exulted in acquiring more and more knowledge.

Working for him was no sinecure. The visits to museums, the dances and parties were interspersed between long, frequently tedious hours of work. When immersed in some current problem, Paul would forget about such trifles as meals and social engagements and one could find oneself, too exhausted to be hungry, still taking notes or discussing some point at two or three o'clock in the morning. He expected me to be available on the telephone, if not in person, twenty-four hours a day, seven days a week. If I was at home in London, it was nothing for him to telephone from Sutton Place at midnight or later and spend an hour or so in a minutely detailed discussion, while still expecting me to be available during normal office hours to cope with any other work that might arise.

A further difficulty with which to cope was Paul's inability to separate business from pleasure; it needed an adaptable temperament and an agile mind to be able to switch in rapid succession, for instance, from discussing the small print of a balance sheet to rushing off to view a painting for sale, then accompanying him on a visit to his dentist, then back for a transatlantic business call, the balance sheet once more – and then the customary mad dash home with half an hour to spare in order to change into long evening dress for a film première!

An additional hazard that not only 'went with the job' but also affected almost every woman who was in any way closely connected with Paul was the bitter jealousy and the intrigue that constantly went on. To the disappointment of some and the disbelief of most, I had not the slightest interest in being one of his 'harem', as they were usually called. As far as I was concerned, he was old enough to be my grandfather and, apart from our working relationship, I looked on him as a sort of adopted elderly relative of whom I was very fond. When we travelled abroad, he acted as my chaperon – and a very strict one he was, too! Paul used to say he regarded me as a mixture of daughter, sister, mother – and father confessor!

We laughed together and we argued. More than once, I hung up the telephone on him in exasperated anger and he shouted at me in a rage. But neither of us ever sulked – one of us always apologised very soon afterwards unless, that is, we had both in the meantime seen the funny side of the whole argument!

Certain subjects, however, remained taboo, as on these we both held widely divergent opinions and, in order not to quarrel, we agreed not even to discuss them. Hence, there are no references in these pages to such topics as racial discrimination, women's liberation, the social services and certain religious matters.

Finally, as anyone who knew him well would agree, Paul was a most complex character: a man of complete opposites, for every quality he had the balancing fault. He caused much happiness and a great deal of sorrow; he envied other people's family life but was unable to feel deeply enough to keep his own together; he regretted his mistakes of the past – and still continued making them all through his life! He was, after all, very human.

As he wished, this book is a remembrance to him and about him, of the good moments, the happy days and the fun I shared with that twinkling-eyed 'Stately Gnome'!

I

Sutton Placed

At ten o'clock on Saturday morning, 10th Otober, 1959, my parents, my grandmother and I were at the front door of our London flat on the point of leaving for our usual week-end at our country house. My telephone began to ring; annoyed, I hesitated – one minute later and we would have been out of the door and away and no one would have answered it.

'Oh well,' I said, 'since we *are* still here, I suppose I had better answer it.'

I picked up the receiver. 'Hello? Yes?' I said, in challenging tones.

'Robina?' came Paul's voice, dubiously.

'Why, yes, Paul!'

I was astonished. He seldom called me on Saturdays as he knew we were nearly always in Kent – and ten o'clock in the morning was an unheard-of hour for him anyway.

'I won't keep you long as I know you'll be off to the country soon,' he said, 'but I just wanted to tell you that I've decided to buy Sutton Place and I've told the Duke of Sutherland that you will be acting for me, and his solicitors should contact you on Monday.'

I was speechless with surprise.

'I know you specialise in company law,' he continued, 'and I want to form a suitable company to buy and run the estate and also to have wide powers – and you can do all that for me and get someone else in your office to do the conveyancing, can't you?'

'Yes,' I said, 'of course I can.'

'That's all, then. I just wanted to tell you, Robina. I won't keep you any longer as I know you'll be off to Kent soon.' And he rang off.

This was a typical example of how Paul could, and frequently did, make amazing pronouncements quite casually as though they were everyday occurrences.

I was then an assistant solicitor with Slaughter & May, the largest English firm of solicitors with offices in the City of London. I think the partner for whom I worked was quite surprised when I walked in on the following Monday morning and told him of the new client I had acquired.

Paul bought Sutton Place from the 5th Duke of Sutherland. It was a tedious transaction as the land had to be registered for the first time and there were never-ending queries and discrepancies to be sorted out – unmarked boundaries, possible rights of way, repairing obligations that no one admitted and so on. In addition, Paul kept on buying a few more acres or another cottage so that the estate seemed to be perpetually increasing in size.

The Duke, who we understood was short of capital at the time, used to get very impatient with what he considered the lack of progress and would every so often telephone to enquire, 'What is holding things up *now*?'

Paul, when the calls were announced, would raise his eyes to Heaven, sigh deeply and spend the entire call – not listening to the testy complaints – but continuing to read his mail and reports until the first gap of silence came when he would rapidly interject, 'Well, thank you for calling, it was real nice to speak with you. I have enjoyed our chat,' and firmly hang up.

Eventually, however, all was completed and, in view of the complexities, in record time.

An incidental story which arose out of this work amused Paul for a long time afterwards. When I was first articled – that is, began my five years of legal studies in order to qualify – a lawyer friend of ours wagered me that I would never finish my articles and qualify but get married long before and give it all up. I accepted the bet – and won!

The same friend then declared, 'Oh, well, I agree you won that – but you won't practise as a solicitor.'

'How much do you bet me?' I asked.

He pondered. 'I'll bet you the first fee you're paid by a client who has come to you personally and not to your firm.'

'Agreed,' I said.

Now, Paul was not the first personal client I had had since beginning to practice – but the Sutton Place purchase was the first matter to be completed and for which the account was rendered and paid. The charges comprised those for the company work and the scale fees for the conveyancing, which are obligatory charges settled by Parliament and based on the value of the property in the transaction; thus, the total bill was not inconsiderable!

I sent (with Paul's amused consent) a copy of the account to our friend, reminding him of his second wager with me. The poor man nearly collapsed! Obviously he had never envisaged anything of that size when cheerfully suggesting he would 'double up' on my first fee – and I firmly insisted that fifty pounds would be an ample substitute.

'Well?' Paul enquired with a grin when I next saw him, 'did you get another nice big cheque?'

'No,' I replied, 'I couldn't expect him to pay all that – my conscience wouldn't let me accept it, so I settled for fifty pounds.'

Paul shook his head. 'If you want to get on in business, Robina, you'll have to get a great deal more ruthless. You'll never get rich that way. After all, it was a firm agreement you had and it was your friend who chose the amount of the wager, so he should have paid up.'

I looked at him intently. 'Even if I had bankrupted him?' (Which it would not have done, in fact.)

'Certainly,' said Paul. 'To you, in a profession, waiving your dues may be a question of conscience – in a business man, it's proof of weakness.'

However, to return to the background of Paul's purchase of Sutton Place, although he came to love the house, he was not at all impressed by it at first. He had been there on previous visits to England but did not seem to have any particular interest in it on those occasions. He went to a dinner party there about a month after he arrived in England in 1959 and was then told by the Duke of Sutherland that the house and estate were for sale. In August, Paul returned again and 'viewed' the house in more detail.

He was much taken with the location and the gardens but afterwards remarked that the dark panelling contributed to a generally depressing aura and that one wing was 'only two large rooms' (the library and the Long Gallery) and half of the adjoining wing consisted of 'a very gloomy hall two stories high' – the Great Hall. These three rooms ultimately became his pride and joy, housing some of the best of his collection of paintings, including some by Rubens, Rembrandt, Titian and Reynolds, together with some fine carpets and furniture.

Paul had at that time a knowledge of art, antiques and antiquities that was very specialised yet, oddly enough, for all his collecting, he did not seem to have an instinctive 'feel' for objects outside his ken. He would work amazingly hard to acquire his learning by reading, going to museums – especially the British Museum – and by consulting the experts. Thus it was that he dismissed the Jacobean furniture in the Long Gallery and the Orussels tapestries there and in the dining-room as 'very poor'. When he later agreed to buy the tapestries from the Duke, he simply would not believe me when I said he had got a marvellous bargain but he felt that he had been over-generous in paying as much as he had. When, a year or two later, they were re-valued, he discovered their worth was then considered by the experts to be about four times what he had paid for them!

On his next visit to Sutton Place, the electric gates at the London Road entrance stuck and he remarked that it did not exactly endear the place to him to have to take such a long drive around the perimeter of the estate in order to be able to get in by the electric gates at the Woking side.

However, he recovered sufficiently from his annoyance to instruct the estate agent, Dudley Delevigne, on the following day to accept the Duke's price for the house and sixty-acre park, saying that, even with its drawbacks, he preferred it to the other properties which had been proffered for his consideration.

When the press eventually found out, it made the front-page headlines in all the English newspapers and, although outwardly Paul complained about all the 'undue excitement' and added he couldn't see what everyone was so worked-up about because, after all, it was not as if he had agreed to buy Buckingham Palace or Blenheim, he was beneath it all thrilled at the interest and read all the newspapers avidly, carefully keeping every cutting.

He started going down to Sutton Place by train as it was so much quicker than by car (Waterloo to Woking was only twenty-five minutes) and, of course, cheaper too: 7/6d. – second class, as he saw no reason to pay 11/- or so, to travel on virtually identical seats in the first class!

It was now that Paul began to assess in some detail the 'credits' and 'debits' of Sutton Place and would recount them to me when we met or spoke on the telephone.

'Cluttons say the floors seem good and the Duke had the electric wiring re-done recently; the roof is fine but the guttering is not, and I guess I don't care for the English habit of no central heating to speak of.'

'Never mind,' I consoled him, 'at least there have been a lot of improvements since the turn of the century!'

He had related, with a mixture of awe, horror and disbelief, that a previous resident of those days had told him that there had been no heating, except for the widely spaced fire-places, no electricity and only one bathroom.

Howards, the builders, did excellent and trojan work for the next few months, stripping and restoring

23

much of the 'gloomy' panelling, putting in strengthening joists and a false ceiling in the Library, modernising and almost re-building the kitchen quarters and a hundred other major and minor tasks as well.

Paul was enormously excited by some of the 'finds' made at Sutton Place while the repairs and redecorations were being done. Behind part of the panelling in the Long Gallery a priest's hole was found, which had been constructed in the days when Sutton Place was owned by the Catholic Weston family and when the English were going through one of their 'no popery' phases. Paul delighted in showing friends how cleverly it was concealed and then, once revealed, the way it had been constructed out of a chimney with room (but not much) for more than one priest to hide.

Another find was the original bread oven behind the drawing-room fire-place. We knew from the old plans of the house that the drawing-room had originally been the kitchen of the house and, sensibly, from the point of view of heating, placed immediately below the principal bedroom. The 'modern' fire-place was thought to have been built about the middle of last century so everyone naturally expected the remains of something older to be concealed behind it.

Paul was almost breathless with excitement when, climbing into the cavity, he announced, 'Why, there's not only the old oven here but bacon hooks as well! Fancy, they've been there for over four hundred years!'

Later, with characteristic thoroughness, he went off to pore over reference books in order to discover what the typical food for that sort of household would have been in the Tudor era.

Meanwhile, the atmosphere around the house became more and more charged and hectic the closer 30th June, 1960, came, as this was to be the night of the 'Great Party'. At one stage it seemed as though the house would never be ready in time and, although the work was not completed, the remaining scaffolding was cleverly concealed, the last-minute paint work just dry enough and the odd step-ladders, tarpaulins and paints hastily tucked well away out of sight. It was not until the day before the party that Paul was actually able to move into the house.

So much has already been written about that particular night's festivities, suffice it to say that what had begun as a coming-out dance for Jeannette Constable-Maxwell, who was the daughter of an old friend of Paul's and a débutante that year, grew into a mammoth house-warming celebration for Paul.

It was certainly a newspaper headline affair and, judging by the late hour in the morning at which guests were still departing, it was a great success.

When the time came for the accounts to be settled, Paul, who always wanted to know the smallest details about everything, was much more interested in what had actually been consumed by the guests than the amount of the bill, so he asked for an itemised account. Having perused it at some length, he at last observed, 'We must have had some fairly alcoholic guests here – I hope that some of them were sober enough to drive safely!'

Twelve hundred guests, arriving after dinner from ten o'clock onwards, had consumed thirty-four bottles of vodka, thirty-nine bottles of gin, fifty-four bottles of brandy, one hundred and seventy-four bottles of whisky, apart from several hundred bottles of beer, Coca-Cola and other soft drinks. They also, incidentally,

ally, ate nearly one hundred pounds of Beluga caviar.

That some of the guests, at any rate, were either drunk or deliberately destructive was evident from one aftermath of the party. The following day, while clearing up the mess and checking the damage – which was considerable – the lodge keepers at the Woking entrance to the estate reported that the electric gates had been broken. These gates, for security reasons, had been closed for the evening of the party and the directions to the guests instructed them to use the London Road (A.3) gates which were kept open all night, so we were very puzzled as to who should even try to go out by the Woking gates. The gates were opened electrically from inside one of the lodges and, by forcing them open, the guests had completely jammed the mechanism which was complicated and costly to repair. Paul could not believe any of his friends would behave in such a way and Jeannette Constable-Maxwell naturally felt the same.

We knew that there had been many gate-crashers who had got in, in spite of the security precautions, under one guise or another. There had been a 'black market' in invitation cards; at one stage, we were informed that people who could not come to the party had nevertheless accepted the invitation and were selling their 'tickets' for up to £100 a person, an invitation for two being worth proportionately more, sometimes £250. Some of the guests who did come brought friends with them who had not been invited and 'smuggled' them in among their group by showing several invitation cards at the entrance gates; with so many guests arriving, there simply was not time to match each card to an individual. Others, in evening dress, crawled through the hedges round the estate and calmly sauntered over the lawns and into the house. Many of Paul's friends who were invited had what I – and he – considered the cheek to ask if they could have a number of extra invitations in order to bring friends of their own. Paul found it embarrassing to say 'No' to some of these requests, because of the people who made them, but managed to curtail numbers somewhat by insisting that no blank invitations were to be used and names of guests must be provided.

So we decided that the damaged gates must have been the work of some ill-mannered, drunken unknown and no more could be done about it.

Several weeks later, Paul and I went to a party given by Maggi Nolan, who was an American journalist then living in London and who must certainly have been one of the gayest and most welcoming hostesses in those days.

The evening was fairly far on when I found myself being addressed by a rather inebriated, plump young man. After the usual inane preliminaries, he nodded his head towards Paul and enquired if I knew him.

'Yes,' I answered cautiously.

'So do I,' he confirmed. 'I went to that party he threw – frightful scrum and all that; bit surprised we didn't have champagne to drink but the caviar wasn't bad – and we had a good giggle at the end.'

Remembering that an unknown hand had removed a valuable antique sugar sifter which later turned up in a London telephone box, I wondered whether this was the 'prankster'.

'How was that?' I enquired, with genuine interest.

He related how, on leaving the party, he and his friends had taken the wrong turning into the drive and found themselves at some white gates which would not open until five of them had pushed and pulled at them for a quarter of an hour or so.

When asked why, as the gates had obviously been locked, no one had thought of going back and finding the right way out of the estate – through unlocked gates – he merely shrugged his shoulders and said that they could not be bothered to drive 'all that way back'.

Gritting my teeth, I pointed out to him that the gates which he and his friends had broken were electronically controlled and were shut on purpose for security reasons – and what did he intend to do about the damage to them?

His bumptious self-confidence suddenly evaporated and his alcoholic haze cleared enough for him to focus a very startled and worried look at me. He did not look any happier when I continued by informing him that I was Paul's English lawyer – and a director of the company that owned Sutton Place and which had had to pay for the costly repairs!

He spent the rest of the evening trying to avoid both of us but, as I had told him I would, I related the story to Paul, who scrutinized him with a cold eye.

Eventually, they came face to face, the young man very ill at ease, Paul fixing him with one of his penetrating stares.

'I gather,' said Paul slowly, 'that you were literally a gate-crasher at my party. Since you are indebted to me for the cost of the damage, I could, and should, send you the bill – but I will come to an arrangement with you: I won't send it provided neither you nor your friends ever come to Sutton Place again,' and, turning away, he took my arm and firmly guided me to the farthest side of the room.

Sutton Place, the English residence of J. Paul Getty

The author's parents, Sir Thomas and Lady Lund, with J. Paul Getty
at Sutton Place

J. Paul Getty in expansive mood

It was heart-breaking to see the amount of damage done during the evening and quite appalling to think that guests at a private party should be so careless and unappreciative both of their host's property and of the house itself. The Library, which is about 140 feet long, had walls covered from ceiling to wainscoting with new velvet panels; ice cream had been spilled on it and, in three places, it had jagged tears, presumably from chairs carelessly pushed back. Several large panels had to be replaced and, although the same colour, they were not from the same vat dye and the difference showed.

In spite of scores of ash trays everywhere, cigarettes were left to burn on tables, window-sills, and the ledges of the beautiful panelling; some of the burn marks were never able to be removed. Sticky drinks were spilled on the tapestries and broken glass left to be trodden in the carpets.

Paul never had such a large party again – the biggest ones afterwards being fund-raising affairs for charity at which people were a great deal better-behaved.

After this, and at the age of sixty-seven, Paul's life as the new owner of a stately and historic home gradually settled into something vaguely resembling a routine. His own movements were always unpredictable but the builders remained in semi-residence for several more months while Paul had further alterations and restoration carried out which, once he had moved into the house, he felt would improve it or his own comfort.

For example, he had a room on the first floor converted into a kitchenette, where he used to amuse himself, particularly on the cook's days off, by indulging in some of his favourite cooking. He was a very good buckwheat-pancake maker and couls also cook a very creditable waffle whih he ocasionally produced for honoured friends.

Although there was a very beautiful outdoor swimming pool, the English weather did not always encourage one to have a swim and, as this was one sport on which Paul was very keen, he decided to build an indoor heated pool, which became very popular with some of his friends and neighbours.

Paul had the walls of the indoor pool covered with Italian marble and was greatly pleased with the effect, although I must admit that I found the pool as a whole singularly unimaginative in design, particularly in view of the financial resources he had available in order to create something comparable with Randolph Hearst's pools at San Simeon, which he often quoted, almost reverently, as being the focal points of that estate.

However, he was happy – and ten times more so when he discovered that, in the pattern of two panels of the marble, there had appeared a sort of Neptune-like face of a man, to which he referred as 'a face from the primeval past'.

'It's quite remarkable to think that that face was formed by the geological movements of the earth long before Man or his forebears had evolved,' he would ponder. 'It's almost as though someone had drawn a blueprint of what was to come.'

When he first moved to Sutton Place, many charities wrote asking if they could use the house for fund-raising events. We were inundated with requests too innumerable for all to be accommodated so we had the unenviable task of choosing some rather than others.

One function to which Paul had agreed was a concert to be given in the Long Gallery, which was on the first floor and about 150 feet long; he was rather dubious about the acoustics and whether there would be any echo. In the early 1960s, one did not need to be as security conscious as, sadly, we have all had to be in recent years and the night-watchman did not start his rounds of the house until after Paul and any guests had retired for the night. Thus it was that, after dinner one evening, Paul suggested we went up to the Long Gallery and, while no one was around, tested the sound-effects for ourselves.

He stood at the top of the staircase looking down the Gallery while I waited at the far end by the two grand pianos, calling, singing and whistling to each other. And that was how, the (allegedly) Richest Man in the World came to be shouting at the top of his voice in the lovely sixteenth-century house deep in the English countryside, somewhere near midnight: 'Beware, lion loo-ooose, 'ware lion loo-oose.'

With the house decorations and furnishings virtually complete, Paul was able to start entertaining and having friends from this country and abroad to stay. There were scores of people who knew him but who had not been to the 'house-warming' and who were, understandably, overcome with curiosity to see his new and much-publicised home. For a while, Sutton Place seemed proverbially like Grand Central Station and, in the first months after he moved in and before he started to count the cost, Paul was an almost incessant and very open-handed host.

This was a new experience for him as up until now, from before the failure of his last marriage, he had spent his life travelling through Europe and the

Middle East, in many respects, I used to think, like a latter-day Flying Dutchman. Of course, he thoroughly enjoyed travelling but as he grew older he became less inclined to make the effort to pack his bags and move again. The trouble was that he had forgotten what living a settled existence in an orderly household was like and he did not really think out the consequences of how his life would change.

When he first moved in to Sutton Place, for instance, he used to get impatient because he couldn't have his meals at any time, on a moment's notice when he felt inclined, as one can do in a good hotel. He would get most irritable at being interrupted, as he considered it, in the middle of his morning's work – which almost never began before half past eleven in the morning – by the announcement at a quarter past one that 'luncheon was served'. He had to learn that the butler was not going to appear at all hours of the night with drinks and snacks – which was one reason why he had his private kitchenette fitted up.

Paul also did not realise, although he was warned repeatedly, that living twenty-five miles outside London was not going to be the same as staying at the Ritz Hotel in Piccadilly, as far as friends dropping in to see him were concerned. He loved to have short visits from people – a drink and gossip for half-an-hour with several people at different times of the day suited him perfectly. He did not care as much for London-based friends making four-hour visits, even though he should have understood that no one wants to make a fifty-mile trip to see someone for just half-an-hour. When we suggested that he came up to London to see his friends instead, he often said, 'I can't be bothered to make all that effort to see someone just for an hour or so. I don't mind making a day of it so much, though.'

'Precisely, Paul,' I would point out to him, 'that is just how *other* people feel about coming to visit you. They would like to but you have put yourself out of their reach – they haven't moved away from you, after all.'

His passionate interest in Sutton Place inevitably cooled after perhaps two years or so. He still admired the house but had grown acclimatized to it and no longer found it awe-inspiring – a real example of familiarity breeding – perhaps not contempt – but a sort of reaction against his previous infatuation with it.

In the 1960s, he was pursuing a very active social life as well as getting through several hours of paper work and interminable overseas telephone calls every day; as he grew older, he became more nervous about being driven back down the A.3 to Sutton Place late at night or in the dark. He became more security-conscious and, as he also became more unwilling to stay in hotels, he either had to stay with friends or to return to Sutton Place by the early evening. The result was that over the years his trips to London became less and less frequent and gradually he isolated himself.

He once admitted that he had made a mistake in buying Sutton Place. Life had not turned out as he had thought it would and he had lost some geographically-distant friends whom he would like to have kept and acquired others whom he would happily have done without. He did not like being 'anchored' to one place; he felt it made him too vulnerable particularly as it provided a static address for begging letters, 'crank' callers, hangers-on and business con-men and, latterly,

33

not only prospecting burglars but prospective kid-nappers.

When Paul had house guests to stay, he was always very generous about allowing them to bring their dogs with them. My mother's Pekingese, Impey, was always welcomed, although we took great care to keep him well away from the Alsatian guard dogs as he, like most Pekingese, would have attacked them and, due to his smaller size, could have got badly hurt in the process. Paul would tell my mother to order whatever food she wanted for Impey from Francis Bulli-more, the incomparable butler. He would produce the most beautiful meals prepared by Kathy, the cook. Normally, we would take these upstairs and feed Impey in the privacy of our rooms. However, one day, we were all sitting in the drawing-room when Bulli-more came in to ask if Impey would like to have his dinner.

'Yes,' said Paul, 'bring it in here.'

So a few minutes later, in came Bullimore carrying a generous bowl of chopped-up food, a napkin over his arm; he carefully placed the napkin on the good carpet and in the centre put the bowl; then he stood back, regarded the dog (who was watching all this preparation with great interest) with a straight face and twinkling eyes and announced, 'Dinner is served, sir.'

Impey, who has always had the perfect manners one expects from the prize-winning Pekingese that he was in his youth, set to and finished it all up with obvious relish without spilling a morsel either on to the carpet or the napkin. Paul, used to the untidier eating habits of the Alsatians, marvelled at the lack of any mess.

34

Turning to the feeding of the human guests, one thing for which Sutton Place was renowned while Paul lived there was the beautiful dining table that was always laid. The 'table' in fact consisted of two very long sixteenth-century tables placed end to end, both of which had come from the late Randolph Hearst's property in Wales, St Donat's Castle.

Bullimore had supervised their polishing until they both looked and felt superb, the dark wood reflecting anything that was placed on it. Consequently, table-cloths were never used but the places would be set with large, protective table mats. The gold plate was only used for special occasions. Normally, the decoration of the table consisted of exquisite displays of flowers from the garden which were arranged in silver bowls and dishes, all reflected in their vibrant colours in the polished oak.

The flower arrangements, which always looked so beautiful, were all done by Frank Parkes (known as John) the footman, and I am sure Constance Spry would have been impressed by his superb creations. John, himself, would be the first to insist on sharing any compliments with Mr Newman, the Head Gardener, inside whose greenhouses one could easily dream hours away amid an incredible display of breath-takingly beautiful plants and flowers.

Many members of the press from all over the world used to visit Sutton Place.

'I seem,' Paul groaned, 'to have become even more newsworthy for buying a stately home in England than I was a newly discovered billionaire by *Fortune* magazine – and that created enough havoc in my life!'

He was very good with them all, in truth thoroughly enjoying the excitement, and patiently agreed to be interviewed, recorded, photographed and televised. He preferred it if the interest was in the house rather than himself, so he was delighted when Radio-Télévision-Française asked if they could make a film about living at Sutton Place, including just a few shots or short interviews with him.

It so happened that I provided the 'background' music to this film. The editorial and interviewing team arrived at Sutton Place with what seemed dozens of technicians in tow and lights, cameras, cables and generating equipment. We all strolled through the house and gardens while they decided what, how and when they were going to film over the next two or three days.

When we came to inspect the Long Gallery, at the far end of which were the two grand pianos, one of the French interviewers asked, 'Do you play these pianos, Mr Getty?'

'No,' said Paul, 'Miss Lund does,' and, turning to me, insisted, 'Come along, Robina, you can play something for us.'

He firmly walked down the gallery, opened up the piano I usually played and sat down in his favourite chair. There seemed to be nothing to do but to follow instructions, so I played some pieces by Schubert, Heller and Schütt for about ten minutes.

When I stopped, the editorial team were nodding at each other and then said, almost in unison, 'There is one piece you play which would be perfect as an introduction to the film. May we record you?'

Before I could answer, Paul had already said, 'That's a very good idea; you can record it in the morning.'

So we did.

36

When a copy of the completed film was sent to Paul and we ran it on his projector, he was thrilled with some of the superb views taken of the house and grounds – and he was as excited as I was as at hearing my playing the 'Canzonetta' by Edward Schütt, which was one of his favourite pieces of piano music.

2

Social - In the Manner of Money

It was not until I began to work for Paul that I saw the 'other' side of some of his friends and acquaintances whom my parents and I had only previously met from time to time socially with him. I was amazed and appalled at how people could behave. Well-dressed and bejewelled ladies who were the centre of attention at cocktail parties would turn up, self-invited, at Sutton Place some days later and, having ostensibly come for lunch, stay on and on until they could get Paul on his own, when they would promptly ask him for the 'loan' of the money to pay perhaps for their forthcoming air fare, the electricity bill or a creditor who was pressing too hard.

He knew, of course, which ones were likely to do this and their 'visits' would often develop into a farcical game of hunter and hunted with Paul, feeling safe as long as a third party was around, using every subterfuge to make sure he was on no account left alone with the lady, and the lady, desperate to be alone with him, using every wile to get rid of the unwanted company! It was not that Paul was actually unwilling to give the money, but he found the presumption that he would 'cough up', as he put it, irritating and the method of doing it distasteful.

'I would much rather,' he often said, 'that someone wanting money called me on the telephone and said so, which would save us both several hours of embarrassment and annoyance. And I do wish people would not ask for a *loan* when we both know they are never going to pay it back.'

Men, of course, also asked for money – not so frequently, but the amounts were considerably larger, running into four or more figures, and such things as outstanding debts or 'temporary financial embarrassments' were never mentioned. Their monetary needs were always wrapped up in tales of marvellous new business ventures in which Paul might care to invest or 'friends' who had some absolutely priceless painting for sale which, surprise, surprise, Paul could have at a special price provided he agreed to buy within fourteen days or so! He saw through this type of camouflage quite easily and the gentlemen usually left empty-handed.

Another variation of the impositions inflicted by some friends was the 'Can we drop in for a drink before lunch?' routine. I must say here that there were friends who did invite themselves without in any way imposing and Paul was always delighted to have them. With others, however, he was used as a method for them to entertain their own house guests cheaply.

The system, when expertly practised, consisted of the hostess usually telephoning on a Sunday morning. 'Paul, darling,' she would say, 'we've got some guests for the week-end – you remember Mimi and Charles, don't you? – and we're just off to church. We'll be passing your gates on the way back and I thought it would be *so* lovely to see you. Can we just drop in for ten minutes for a teeny drink?'

Paul, naturally, would say yes and add that perhaps they would like to stay for lunch. That, being the original intention, was accepted with grateful alacrity.

The inference having been that there would be just an extra four guests for lunch, the arrival of the party would reveal that Mimi and Charles were only two out of perhaps eight or ten guests who were apparently staying for the week-end. So that would be a free lunch for about a dozen people.

After lunch, the hostess would ask Paul if they could all *possibly* see round the house, and his collection of furniture, carpets, paintings and tapestries would be exclaimed over. If the 'tour' of the house appeared to be ending too soon, there was always the garden to be inspected. With any luck, this would use up the afternoon until five o'clock, when a delicious tea of sandwiches, home-baked scones and cakes was always produced by Kathy, the superlative cook. This would last until about, a quarter to six when a skilful hostess would remember that Paul had not shown her guests the outdoor – or indoor – swimming pool. Once this was organised, someone was bound to remark how lovely it would be to have a quick dip in the heated water – and so swimsuits were provided. After the swim, it was back to the house for a drink and to say good-bye – just as Bullimore the butler was enquiring how many were staying for dinner – so, of course, they would stay on!

Eventually, such parties would leave at ten o'clock or so at night, the skilful hostess having provided a day's meals and entertainment for her delighted guests at no cost to herself.

Paul was no one's fool and fully realised what was happening but tended to shrug his shoulders. Sometimes he would find one of the guests particularly interesting or amusing and would monopolise his company; if he was bored, on the other hand, he would just walk off to his study, muttering that he had to make a telephone call or do some urgent work and would not re-appear until the next drink – or meal – time.

'Why do you put up with people who use you?' I sometimes asked him.

'Because they nearly all do. If I gave up seeing everyone who uses me or has taken advantage in some way, I wouldn't have many friends left – and, in the end, one must have someone for company to talk to. Most people are out for what they can get, you know.'

The sad part was that he was such an appalling psychologist that he was quite unable to distinguish between the 'friends' who did use him and those who wanted to give, not take. The ultimate result was inevitable. The most genuine friends whom he could have had were often quiet, shy and anything but pushing and, consequently, did not presume to invite themselves to Sutton Place. Paul would misinterpret their motives by deciding that they were not very interested in him and the potential friendship would gradually fade away. He was flattered, on the other hand, by the attentions of the gushing types who were always keen to come to see him – *even* though he knew he was being imposed upon.

He was occasionally moved to remark about some of them that it would be the day when he was invited back. It was, to be fair, a rather daunting prospect to invite a billionaire to a meal, particularly to those many people who did not run an 'establishment' with a large staff. We, as a family, always treated Paul as one of us and we ate the same food and used the same table ware as we did when he was not there on the principle that, if he did not like it, he need not come again – but he always did.

One couple who, on many occasions, had drunk, lunched and dined with Paul, one day invited him to go back with them for a 'light supper'.

The next day, I asked him how he had enjoyed himself.

'Well,' he said, 'firstly, all they had in the larder was spaghetti; next, I found it was tinned and it was only a medium-sized tin for the three of us and, lastly, we couldn't find the tin-opener for an hour, by which time I had been hungry but the feeling was wearing off. I came home soon after and had some roast beef sandwiches and a large slab of chocolate. Then I felt better.'

Some of the guests who stayed at Sutton Place left their manners behind them, if they ever had any, and on occasions their behaviour was little short of disgusting.

It was quite common in the first years after Paul moved into Sutton Place for guests, soon after their arrival, to go round replenishing their own cigarette cases from the boxes that were, naturally, left out for normal use. This would sometimes happen even when the guests had only come for a drink or Sunday tea. One day, I had noticed all the cigarette boxes being filled up before the arrival of a party of 'friends'. When the six of them left, two hours later, there were ten cigarettes left out of over two hundred that were put out. When one considers that this frequently happened, I don't think it particularly surprising that eventually Paul instructed that cigarettes would only be put out on certain specified occasions.

Another similar situation often happened with the drinks cupboard. This was a useful little mirror-lined cupboard in the television room, which was next door to the drawing-room. Normally, one would have expected guests to wait to be invited to have a drink but, as people found out where the bottles, glasses and other accoutrements were kept, they frequently helped themselves (very generously) and, on more than one occasion, I found someone – usually a friend brought by a friend of Paul's – filling up his own hip flask with Paul's whisky or gin. Once, when I was so indignant that I remonstrated quietly but firmly with the man, he simply shrugged his shoulders and said, 'The old so-and-so can afford it!'

Another house guest was a fairly hefty wine drinker – and looked it – and never found the amount of wine supplied at meals enough for her requirements so she used to take bottles up to her bedroom. After she left, the staff would go round the room she had been in finding empty bottles behind the curtains, at the bottom of the wardrobe and even hidden in the fire-grate.

During one visit of hers, the house suddenly seemed to develop a plague of mice – up on the first floor. We later discovered why: apart from the wine bottles, which presumably were of no special interest to mice, there were half-eaten bread rolls and bits of cheese and cake hidden away in drawers and cupboards.

Although Paul's own social manners frequently left a lot to be desired, he had double standards and was very aware of other people's short-comings in that direction. He would often sit down at a table before any of the ladies had even had their chairs pulled out for them and he frequently ordered his own meal, even when he was host, before anyone else had a chance to do so themselves. He often walked through a door ahead of a female companion, but would criticise a younger man for not, in his turn, holding it open.

He was, however, very correct in some matters and as meticulous in giving formal invitations, making sure that, if any friends of his who were invited to Sutton Place themselves had house-guests, then those guests were also to be included in the invitation. He also took a surprising amount of trouble on purely social occasions to gather people of some common interest together.

He disliked invitations from hostesses whom he thought were 'social-climbing' and would not invite them back. He was quite aware that he was often used as an attraction ('a modern version of the old fair-ground side-show', he would whisper to me) at an otherwise unremarkable or downright dull dinner-party or other social affair; but, if for one reason or another, it amused him to go, he would and let his hostess preen herself on her success in making such a catch.

If he felt that someone had committed a social gaffe or been what he considered ill-mannered, he would not remark on it at the time but he never forgot and would always thereafter maintain a cool attitude towards the offender. A small example was the annoyance he felt for the Principal of a well-known London finishing school. She used to invite Paul to attend the school's annual ball, which he quite enjoyed.

He once remarked, 'There are a lot of very attractive young ladies there, so any man would enjoy himself when they seem to be brought up and presented in a never-ending procession – and it doesn't tax the mind too much as I only have to make the conventional responses over and over again! Mind you,' he added mischievously, 'the fun comes when one makes an unconventional answer; they haven't been taught how to cope with that, poor dears, and they tend to look a bit dumb-founded sometimes!'

On the occasion in question, I took Paul as my guest to a white-tie-and-tails affair at Guildhall. We had just moved away from being received when Paul was 'homed on' by a short, white-clad, plump woman. He shook hands with her and introduced us. After a quick handshake and a quicker smile at me, she turned back purposefully to Paul and, in a voice whose clarity must have extended over many square yards of that large reception hall, addressed him (he said afterwards, rather as though he was one of her pupils), 'Now you *will* come to our next Ball, won't you, Mr Getty? There will be lots of pretty gels there and we'll have plenty of really charming dancing partners for you. My gels are such attractive and sweet things this year.'

'Thank you, I shall be pleased to come,' and he bowed slightly, smiled a little stiffly, took my arm and very firmly moved me on.

Once we were out of earshot, he gave vent to an indignant snort and a scowl. 'Well,' he remarked sharply, 'my mother would never have done that! I was always taught by her that you never gave an invitation to someone in anybody else's presence without including them, too – otherwise you kept quiet until you got your intended guest alone.'

He did go to the Ball, as promised, but it was the last time and when one of the 'young ladies' from the finishing school was subsequently very badly behaved at Sutton Place, he remarked with some asperity, 'Between her mother who is a typical social-climber and her 'finishing' which omitted to teach her good manners, I'm not surprised. She needs a good spanking from her father – but then he's too indulgent to do anything.'

My parents and I were often asked to act as 'locum' hosts when we stayed at Sutton Place, looking after his guests when he did one of his disappearing acts. Often the guests that he thought were going to bore him, turned out to be delightful and interesting people. We always felt sorry for those who had never been to Sutton Place before and were probably terrified by the comparative 'formality' of it – a reaction that Paul never had himself and seemed unable to comprehend.

I did not altogether blame him for trying to avoid some self-invited arrivals who came with purpose glinting in their eyes. There was a certain American gentleman, to whom Paul referred as the 'Chief Scout for Finances' for one of the American universities, who seemed to make an annual European tour, visiting all the wealthy American expatriates, soliciting funds for this or that university project.

On this particular occasion, he arrived for a pre-lunch drink with his wife and daughter. The daughter seemed a nice girl but hardly uttered a word; the wife was one of those women who wants to be taken for her daughter's younger sister. She had startlingly black long hair, carefully arranged in casual style over her shoulders, the longest black false eyelashes I have ever seen, and she wore very high, stiletto-heeled bootees because, as she explained confidentially to my mother, 'I have weak ankles and can't wear shoes.' (The logic of that explanation still eludes us!)

Quite naturally, our American gentleman didn't want to discuss his business with all of us around and probably he felt it inappropriate at the luncheon table anyway. Paul was quite aware of this, and not wanting to be cornered into being asked for money he had no intention of giving, he suddenly left the table just before the coffee was served, with the usual excuse of a telephone call.

However, he was outsmarted by the American family who, their mission still not having been accomplished, stayed on and on for tea then drinks until, once again, we were all seated at the dining table.

As it happened, dinner was the turning point due to an entirely fortuitous occurrence. My mother was sitting on Paul's right, the American wife on his left; her husband was next to my mother and I sat on his other side thus diagonally opposite his wife, who was next to my father.

The conversation was continuing quite happily on non-controversial topics, the soup was brought in and, for a few moments, most of us concentrated, eyes down, on enjoying the vegetable soup. When I next looked up, I became suddenly aware of a startling change in the American wife: she had lost the top half of one of her pairs of long, lustrous lashes and the one-eyed effect was rather bizarre. Involuntarily, my eyes dropped to her soup plate, for the lashes must surely have fallen in that. However, without craning my neck too obviously round the flowers in the centre of the table, I could not see very easily, so I cast a surreptitious glance round the table to see if anyone else had noticed and found that Paul was sitting transfixed, staring at her with his soup spoon half-way to his open mouth. Luckily, he pulled himself together and continued with his soup while glancing to see if anyone else had seen anything. When he saw me watching him, he gave an almost imperceptible movement of his head towards the poor lady and slowly closed one eye in a gigantic wink. I nodded – and we both got a fit of the giggles.

We watched every mouthful she took from that plate of soup and the eyelashes never re-appeared. He said afterwards that he had been wondering what he would say if he suddenly spotted them in, say, her spoon: 'Excuse me, but I think those are your false eyelashes,' or 'Excuse me, but I think there's a centipede in your soup.'

In the end, Paul at last invited the American gentleman in to the study for a private conversation and ended up by giving him a donation, though perhaps not as big as had been hoped for!

Paul later remarked, 'I expect he went away thinking that his persuasive turn of phrase induced me to make a contribution to his goddam university. Little does he realise that it was really conscience-money for my laughing at his poor wife's misfortune – and my admiration that she carried on regardless, though I do think,' he added pensively, 'that I would have removed *all* my false lashes if I'd been in her situation!'

Over-familiar ladies and gushing men frequently homed in on him like bees to the hive and, although he usually tried to maintain some social good manners, he used to get very bored with the excessive compliments, aware that such people usually were so busy talking that they seldom bothered to listen to anyone else. He would sometimes reply to the introduction of a particularly effusive individual 'So pleased to have you meet me.' No one ever realised what he had said.

48

'Like most people,' he would say, 'they hear only what they expect to hear, which is why they never learn anything new.'

Although Paul readily accepted some invitations, he dithered over others, frequently changing his mind on the day of the event – usually about two hours before he was due to leave Sutton Place for a London engagement. Where it was a ticket function, such as a film première, charity reception or even a ball, the repercussions were not too bad as he was quite agreeable to the tickets being handed on either to friends or his personal office staff.

The difficulties arose, however, where he was supposed to be going to a private dinner party at which he was to be guest of honour, or a formal 'sit-down' affair, with a printed table plan, where his absence would be only too obvious. I used to expend a great deal of nervous energy and time cajoling, arguing, reasoning and using any other methods I could think of to get him off. The dialogues nearly always followed the same pattern.

RL: 'It's time you got changed into your dinner suit – you have to leave here in half an hour.'

Paul: 'What for?'

RL: 'You're having dinner with Mrs X.'

Paul: 'With Mrs X? Why didn't somebody remind me?'

RL: 'Somebody did. I did, your secretary did and Mrs X telephoned this morning.'

Paul: 'Well, I can't go. What these people don't realise is that I'm an old man and a working business man and that Getty Oil doesn't run itself. It's all very well for a man who has a nine-to-five job, shuts his office door and goes home leaving his responsibilities behind him. I've got a company that has business interests around the world. I'm expected to be at the end of a telephone twenty-four hours a day because you can bet your bottom dollar that somebody in some part of the globe is going to make a goddam fool of himself, the company and me and, since few people seem to have enough sense to come in out of the rain, I have to be around with "the umbrella" . . .'

And so it would continue for some minutes more. When, ultimately, he paused for breath, having covered the inability and unwillingness of employees to take responsibility, coupled with their unworthiness to be given it, though, heaven knew, he would be only too pleased to have some of the burden taken off his own shoulders, I would continue:

RL: 'If you don't want to spare time off work to go to social engagements, then don't accept the invitations in the first place. People won't take offence if you refuse on the grounds of age and business but, having accepted to go, it's extremely bad manners not to bother to turn up, particularly when you are guest of honour and where your hostess has invited others especially to meet you.'

Paul: 'Well, you know' – with some asperity and sarcasm creeping in – 'the other guests will still be there and there should be enough of them to provide some conversation, and I guess the dinner will survive, like Mrs X herself, without me. So you can just telephone and explain I can't go.'

Having arrived at the situation where he adamantly refused to go to his dinner party, or whatever else it might be, and equally adamantly refused to make his own excuses or apologies, there was one last gambit which sometimes worked.

'Right,' I would say, moving towards the telephone, 'I'll ring Mrs X now and explain that you're very sorry that you can't go to her dinner party tonight but you're feeling your age and, as she'll understand, too tired and shaky to be able to make the journey and stay out that late.'

'That's not what I said at all,' Paul would splutter indignantly.

But before he could say any more, I would smile sweetly at him and retort, 'It's what *I'm* saying. If you don't like it, you have two alternatives: either ring yourself with your apologies, or go. Now, which is it?'

Nine times out of ten, with a scowl at me, he would look at his watch and, anything to avoid making that telephone call himself, get up and mutter, 'Well, I suppose I've just got time to get changed.'

As he got to the study door, he would pause, look back and throw the last word at me, 'But it's the last time I'm accepting any more social engagements,' and disappear muttering.

I would look at the rows of invitation cards on the mantelpiece and smile to myself.

After one of these little episodes, Paul would turn up for his evening out, full of charm and good humour and many flattering remarks for his hostess and, provided he met a few other guests who interested him, he would thoroughly enjoy his evening and no further mention would ever be made of the battle to get him there.

As with all rules, there are exceptions, but basically the people whom he liked to meet fell into several fairly easily recognisable categories: if they were women, they had to be at least fairly attractive

(women who were real beauties made him feel un-
comfortably inadequate – 'they attract so many men
around them,' he would say, 'that I find there is too
much competition') or, failing that, have plenty of joie
de vivre. With these attributes, nothing else was necessary; without them, they had to be heiresses, titled or
members of a newsworthy social set.

As for men, he admired those who were handsome
and well-built, but strictly and only in the context of
the 'Mr Universe' competition, otherwise he did not
encourage their company. Other requirements, as far
as Paul was concerned, were money, power or a title,
probably in that order, and preferably all three.

In another class altogether, he had a great deal of
time for professional experts – provided their expertise
was relevant to one of his business or personal interests.
Then, unless those experts were very skilful, Paul
would corner them, even when out socially, and literally wring them dry of all the information he could
squeeze out of them – and not expect to be charged for
it. Once he felt that he had obtained as much en-
lightenment as was possible from a particular source,
then that individual would be dropped and forgotten.

In his moments of candour, Paul would remark, 'I
guess I am like most Americans. Because we come
from what is supposed to be classless democracy, we
go overboard for those whom we see as the aristocracy
or upper class, while at the same time we're a bit
nervous of them. You see, deep down, most of us have,
though we pretend not to, a social inferiority complex
– we're only really socially at ease with others who
are, I suppose, middle-middle or lower-middle class,
and we're uncomfortable with the so-called lower
orders because we don't know how to behave towards

them. That's why so many Americans are aggressive – they don't feel equals but they're not damn well going to show it. And because, socially and culturally, they so often feel inferior to Europeans, they get pushing and thrusting in business because, by George, they're going to come out tops in something!'

Then he would grin, raise his glass of rum and Coca-Cola and toast – 'Bottoms up!'

3

Money and Parsimony

While Paul was a great admirer of John D. Rockefeller III, he looked on him as a sort of dedicated, misguided saint. 'He's a very wonderful person and one of the really great philanthropists,' Paul would say.

Mr Rockefeller tried hard on several occasions to convince Paul that great wealth carried with it a moral obligation to help, in whatever way was possible, those who were less fortunate. Paul, however, was not to be moved.

'I tried to explain to John,' he said, 'that I consider the best thing one can do for people is to provide them with jobs so that they can earn a good wage and look after themselves. I don't believe in the Welfare State – those that deserve it, don't get it, and the lazy, good-for-nothing scroungers are encouraged in their idleness. People only really appreciate what they pay for: if it's free, they waste it. If you have to struggle for something, whatever it is, you appreciate it when you get it, so you value it and take care of it. The current fashionable type of socialism merely encourages people to believe they are entitled to whatever they want, then they get selfish and want to take everything but never dream of giving back in exchange – or even voluntarily.'

These views did not go down at all well with people asking for money, so Paul seldom mentioned the true reason for his cautious approach to the question of charity as a whole. Of course, he gave many gifts and donations to organisations and individuals but with the proviso that no publicity was to be given to him as this would merely encourage even more people to write begging letters to him. The amounts he gave were usually modest in view of his wealth but, on the other hand, the demands were many – and most were turned down, quite justifiably, on the grounds that he could not possibly help everybody who asked for it. To friends who tried to inveigle him into some commitment, he would just reply, 'I'm too poor to manage that, particularly in view of my prior commitments,' and refuse to budge an inch.

The number of begging letters that came in varied according to the current amount of publicity which Paul was receiving; sometimes it could be as bad as two hundred in one day. Those that Paul did help were by no means always grateful – in fact, far from it in a few cases, where the writer seemed to think that his was the only letter written, the only deserving cause and that Paul had a few million dollars to hand, in cash, available for instantaneous gifts.

On one occasion, he sent a donation of £100 to an armed service charity and received a most abusive letter back from the charity's Secretary, which began on the lines:

Dear Mr Getty, The sum we require and for which I wrote and asked you is £10,000 not £100 which you sent and which is laughable if it were not so disgustingly mean . . .

I was indignant and, noting that the writer had not

returned this 'disgraceful' donation, observed to Paul that he really should write and ask for his money back as he could easily donate it to many other causes where it would be appreciated. He smiled and shook his head.

'The charity is a good cause,' he said. 'Why should it and its dependants suffer from the ill-manners of a petty-minded, rather stupid little man?'

There was a taxi-driver in Vienna who nearly ran over two of the world's richest men – at the same time. Paul had invited John D. Rockefeller III to dinner and the three of us were strolling from the Bristol Hotel to Sachers. The two gentlemen were soon deep in a business conversation and I stopped for a few seconds to do a little window-shopping. When I looked round again, there were the two of them, still completely absorbed in each other, about to step off the kerb right in front of a taxi.

'Hey, wait,' I bellowed and they both stopped and looked round in bewilderment. After that, we concentrated on arriving at Sachers safely!

Once seated at the table, the two millionaires continued their conversation and I contemplated each of them in turn.

If they had been knocked down and had had no means of identification on them, I doubt very much whether anyone would have guessed who or what they were. Mr Rockefeller was wearing a very unobtrusive brown pin-stripe suit with a muted blue and brown striped tie; his brown trilby one could have found in a thousand shops in England and his raincoat looked as though it had seen better days – some time ago! The whole effect was of a total lack of ostentation, as he intended, with his customary modesty; a man you would hardly notice, even less remember.

Paul, on the other hand, was wearing a dark blue suit which was obviously an expensive one but the trousers were too long and were sagging over his shoes and bagging over his knees. I had noticed earlier in the evening that he had a hole in the heel of one sock and I now saw that the collar of his blue and white striped shirt was frayed.

I had to smile to myself at my view of these two charming, powerful, intelligent and modest billionaires, each anything but the usual public conception of the fabulously wealthy.

Paul's disinclination to spend money on his clothes, wearing them until they frayed or shone, often led to a battle of wits.

On one occasion, it was nearly midnight and after a long day's work in the heat of Vienna in the summer, I was dropping with tiredness. I was thankful to retire to my hotel bedroom and sit down in a comfortable armchair for a rest before getting undressed. After about ten minutes, I became aware of a quiet but persistent knocking at the door.

'Who is it?' I called.

'It's me, Robina,' Paul's muffled voice answered. 'Are you awake? I won't keep you a minute.'

'If I had been asleep, I would certainly be awake now,' I pointed out as I opened the door to find him holding a shirt and two pairs of socks.

'Dear, I wonder if you would just turn the cuffs of my shirt as they seem to have frayed a little and my socks have holes in them.'

I looked at him, open-mouthed. I took the shirt: I already knew it – the cuffs were a disgrace but not as bad as the collar. I took the socks: all had at least one hole in the toes; three also had holes in the heels.

'No,' I said, 'absolutely and finally *no*. Legal adviser, personal assistant, press officer, occasional secretary, nurse, navigator, chauffeuse, chief packer, yes; I might even sew on a button or two for you – but darning and turning cuffs, *No*. The only place for those is the rubbish basket.'

'But don't you think I could get some more wear out of the shirt?' he ventured.

'Maybe someone could but not you and the answer remains *No*. *You* can't go round in those old things any longer.'

'Well, I suppose not,' he said doubtfully.

I knew there was one argument which would possibly convince him. 'Your mother would have been horrified to see you wearing these; you, yourself, have often said she was always so meticulous.'

'Ye-es,' he nodded slowly – then, more decisively, 'Yes, Robina, you're quite right, quite right. I'll throw them away now. Good night, dear.'

And he disappeared purposefully down the corridor.

I closed my door and made a mental note to get him to a shirt-maker at the first available opportunity – and I still was not quite convinced that he would actually throw the shirt and socks away.

As it happened, the following day presented the opportunity.

Paul was completely predictable in that he would always behave unpredictably – according to other people's customs, anyway – but it was occasionally possible to play him at his own game.

This particular afternoon, he told me we were going to do a round of the antique shops in Vienna so I fetched my coat, put on some comfortable shoes and

went to collect him. He then decided he would telephone the shops instead so he disappeared and pro-
eeded to ring every single one, re-appearing two hours
later, at ten minutes to five, to say we were not going
as no one had anything he wanted. Instead, he said,
we would go for a walk.

'Fine,' I said, and we strolled out.

I had already found out the addresses of the nearest
good Viennese shirt-maker and gentlemen's outfitters.

I gradually managed to propel him in the right
direction until we were outside a very well-
recommended shirt shop, then turned him round to
face the door.

'I am not going to sit down to dinner with you any
more while you're wearing shirts with frayed collars
and cuffs,' I addressed him in a 'no nonsense' voice,
'so in you go and buy some new ones,' and I pushed
him, gently but firmly, through the door.

Twenty minutes later he emerged, pleased as could
be with two drip-dry shirts and four pairs of socks.

For some reason, Germany always seemed to bring
on excessive fits of thrift in Paul. Once, arriving in
Baden-Baden, I told Paul I was going to telephone my
mother. There was a lot to relate, which I had already
told him. We spoke for sixty-eight minutes, and Paul
came knocking at my door as I rang off.

'How long do you think you've been talking?' he
asked, furious.

I pointed out that I had spent over two hours telling
him what I had spent just over the hour relating to my
mother; that, as he knew, due to her very bad eyesight, my mother could not see well enough to read so
I was unable to write to her with my news; that, as
he also knew, I was not in the habit of making long

calls to her and lastly, as he customarily made £200 to £300 worth of telephone calls in a day, I felt that one call of mine at £15 was not excessive once a month!

'However,' I continued, 'I'm not prepared to have you moaning at me so I will pay for the call.'

With that, I thrust the money into his hand. I realised, of course, that his irritation had been nothing to do with the telephone call as such, or even the cost, but the fact that I had walked away from him and left him on his own. He never bothered about leaving anybody else alone but he could not stand it happening to himself, just as he did not like to be kept waiting (although he, as I have already said, was often late). No more was said then – or, in fact, until years later, when he referred to the episode with a mischievous glint in his eyes as 'the time you nearly bankrupted me!' – but, next day, when we went out for a walk round the shops, he suddenly darted into one saying, 'You wait here a minute,' and emerged shortly afterwards with a small parcel which he thrust into my hand.

'A small remembrance,' he said.

I opened it. It was a beautiful fountain pen, costing, I knew, well over 100 D.M. I thanked him, with genuine appreciation, and he asked, 'Friends again?'

'Of course.'

In 1960 we went to Munich. I was thrilled with my room at the Vier Jahreszeiten Hotel: it was a very attractive room, beautifully decorated. Everything was so concealed it took me ten minutes to find the electric light switch!

The next day, when I had just finished unpacking all the business papers, ordered some coffee and was preparing to sit down and have a short rest, there was an imperative tapping at the door.

. Paul Getty and Robina Lund at her family home in Kent. . . .   Time for lunch . . .

. . . and after lunch

'Yes?' I asked. 'Who is it?'

I opened the door and Paul came in with a determined look. 'I'm sorry, Robina,' he said, frowning, 'but we'll have to change rooms.'

'Oh, no! Why?'

'I've only just noticed the tariff; these rooms are over £10 a day. I've arranged with the Hall Porter that the hotel will move our things down to two more reasonably priced rooms on the second floor. They're very nice but more modest.'

They were quite nice – but about half the size – although it wouldn't have been so bad if I had not had the other gorgeous room first! Our 'more modest' rooms effected a big saving as they were only just over £3 a day – and I expect that was how Paul was encouraged to spend several hundreds of pounds the next day buying pieces of antique furniture.

He remarked a short time before he died that he would be only too pleased to get a hotel room for £10 a day and the idea of any longer being able to stay in a luxury hotel anywhere in the world for £3 a day had become laughable.

As I have said, in another context, many people misjudged Paul by presuming his motives and goals were what *they* thought they should be. His conservation prize was a typical example.

All the talk of philanthropy was, as Paul himself put it, 'eye-wash'. I have already quoted his idea of philanthropy as giving a man work and paying him for it, so that he could look after himself and his family without needing to beg for help.

Conservation of wildlife was of only passing and muted interest to Paul – he was infinitely more concerned about the restoration and preservation of old

c

master paintings, antiques and ancient edifices.

Having been, myself, closely involved with various aspects of conservation over the years, I did my best to try to educate him on basic facts, such as which animals were rare or actually endangered, and the importance to the world climate of the great rain forests, but Paul's will to remember was not there.

In 1974, he instituted the J. Paul Getty Wildlife Conservation Prize, to be awarded to the individual who, in the opinion of an independent panel of experts in 'various fields connected with conservation' had contributed the most towards the objectives of preserving and conserving wild places and animals in order to ensure human survival in an ecologically balanced world. However, to him his conservation prize represented an outlay of approximately $50,000 for perhaps $250,000 or even $1 million worth of publicity for the name Getty and thereby for the business.

Some people have been shocked by his attitude – but at best the prize drew the attention of the public all over the world to the desperate need to conserve what we have not already destroyed, in order to survive at all. It also provided the catalyst for assembling invaluable information as to what projects not only large organisations but hitherto unknown individuals in desolate regions had been undertaking in the field of conservation.

The stories about the 'pay telephone' at Sutton Place have been repeated ad nauseam and almost always without pointing out that many people, guests, visitors and journalists alike, took advantage in Paul's early days there of the many telephones around the house. People who were not close friends of his would make some excuse to get to a telephone and then make a long-distance call to Japan or South Africa, talk for half-an-hour and walk off without even offering to pay. Consequently, it was scarcely surprising that a pay telephone was installed for visitors, other than house guests and close friends, who were always at liberty to make what calls they wanted.

Paul could be just as careful with his own telephone calls as with other people's. The following story has always been one that has made me chuckle – from more than one point of view – as it is a neat illustration of how two wealthy people could react.

One day, the telephone rang in Paul's suite at the Ritz Hotel. Paul answered it. 'Hullo,' he said in his usual slow and cautious way.

'This is the operator: will you accept a reverse charge call from Dunrobin Castle, from the Duke of Sutherland?'

'Well, yes,' he said – then added quickly, 'but not for longer than three minutes.'

Paul's self-inflicted frugality applied to his own enjoyment as well. 'I'd like a trampoline,' he said one day at Sutton Place.

The previous evening we had watched a television programme of trampoline gymnastics and he had been enthralled.

'Then have one,' I encouraged him.

'Where would we put it?'

'Well, you need a good bit of headroom, so what about the unused squash court?'

'It's not in repair.'

'Then repair it.'

'Too expensive.'

'How much?'

'Four thousand pounds, five thousand maybe.'

'So what? You'll have a whole covered building re-tiled, re-pointed and re-glazed. Surely that's a good investment?'

He contemplated. 'No,' he said at last, 'it's not. I don't want to play squash and paying five thousand to bounce on a trampoline *is* extravagant.'

So, no trampoline!

Paul frequently used to borrow the odd pound or two off me; sometimes he remembered to return it, otherwise I just 'borrowed' it back again when he was in funds. On one occasion he did rather well and left me the loser as well as the winner.

We used to go to the Pied Piper's Ball as guests of Vere and Patricia Harmsworth; like most Charity Balls, they always had a draw for Raffle prizes during the evening and very splendid the prizes were, too. Paul was always generous in buying Raffle tickets when I was with him as I was so often lucky in winning. One year, he bought twenty tickets and handed them over to me with the strict admonishment not to leave them lying about or lose them.

This time, I (or we) won first prize – fifty pounds in cash. I then also won third prize – a magnum of champagne, which I handed back to be re-raffled. Paul was thrilled with the cash prize and a little annoyed and puzzled at my handing back the champagne, even when I explained that it did not seem right to me to keep two out of the five prizes.

He preferred to leave most dances well before they finished and end the evening in the quieter surroundings of a nightclub, his favourite being the '400' when it existed.

'Would you like to go to the "400" and celebrate

our winnings?' I asked.

'Oh, yes, that would be fun.'

I knew how much Paul, who was so often expected to pick up the bill, really got a thrill out of a surprise treat, so when we were settled at our table at the '400' I said to him, 'I'm paying.'

'Thank you, dear. Now, let me see,' (putting on his glasses and studying the menu carefully), 'shall we have a bottle of champagne,' (I should never have given back that magnum!), 'and what would you like to eat?'

We had already had a very delicious and ample dinner not so long before and I was certainly not hungry but Paul was going to enjoy his 'special treat', so he ordered soup, avocado, then kedgeree, followed by ice-cream – while I had a modest, but tasty, welsh rarebit and coffee.

When the bill came, I handed him more than the required money under the table so that it wouldn't be seen; having paid, he then slipped the change into his pocket!

By the time we left, it was well after 3 a.m. and, although Paul was staying at the Ritz Hotel, the chauffeur, Lee, had to drive back to Sutton Place. Paul always gave him what he called 'a fine' if he kept Lee out after midnight.

We were on the way to drop me back at my home when Paul leaned over and whispered, 'You might lend me five pounds out of your winnings to give to Lee, dear, as we've kept him up so late.'

'Of course,' I said and handed it over.

I never got it back, needless to say – but then, I hadn't expected to.

He very proudly told several people that I'd made

'a profit of fifty pounds' on the evening; two of his 'lady friends' were sufficiently jealous to remark that it was most unfair that I should get all that money.

My actual 'profit' was twenty pounds!

I sometimes felt like a personal accountant when we lunched or dined out. When we were on our own, and sometimes even when other people were there, he would pass the bill over to me to check the items and to add up the total. If that was correct, he would then enquire whether a service charge was included and, if it was clearly marked down, ask if the menu specified anywhere that it was to be included. He hated being taken for a ride, as he considered it, by some restaurants who made no reference on the menu or bill that the service charge would be included but nevertheless slipped it in – and then added insult to injury by expecting a tip in addition.

He used to get so annoyed about the combination of service charges and tips that he usually said to me, 'You work it out, Robina, and let me know how much I should leave as tip.'

He did not deserve his reputation for mean tipping on these occasions because I used to tell him what my father, who always tips generously, would leave and Paul would put that amount down and then some more.

He had some strange ideas about saving money at Sutton Place and I often teased him on what seemed such illogical economics.

For years the hot water and central heating systems staggered along on an inadequate boiler system because he did not want to spend the money on a new boiler. At one stage, in order to stop the rising costs of the central heating, he had it cut off from the bedrooms and instead put one or two hot air heaters into each room when in use and the heaters would be kept on virtually twenty-four hours a day.

'But Paul,' I would try to reason, 'what you may be saving on the cost of fuel for the central heating you are certainly spending on electricity to run the heaters.'

'No,' he would reply in measured tones, 'I've worked it out carefully and I think I save about fifty pence per room per day – and that adds up, you know, over the winter months.'

Similarly, as the years went by, some of the carpets started to become threadbare – even the one in the drawing-room had a bald patch in it where feet coming in and out of the television room had worn it away, but he refused to spend the money to have the whole room close-carpeted. He only agreed to the carpet in the private office where his personal staff worked being renewed when the holes in it made it positively dangerous and we warned him that, one day soon, someone was going to catch his or her heel in it, fall over and break a limb.

On the other hand, he was quite agreeable for the cloakroom to have new furnishings and wall covering even though he said he thought it a trifle expensive at £1,400!

4

Press - Stop and Go

One always had to be prepared for the unexpected with Paul, especially as, not ever having been concerned with running his own home or doing anything for himself, he often did not realise the adjustments one had to make to accommodate his sometimes unorthodox manners and actions.

Soon after he arrived in England in May, 1959, we invited him down for the first time to spend the weekend with us in Kent. Being four in the family, and our house being of the old, Kent smuggler's type and thus not very large, we arranged for him to stay at the very excellent country hotel at the local town three miles away. As our house was in the middle of country roads and none too easy to find, we told Paul to drive direct to the hotel and we would meet him there and then take him back home.

He arrived on the Saturday morning, slightly late as usual, and we were a little surprised to see not only himself and the chauffeur, but another man climb out of the car as well.

After the first greetings were over, Paul waved a hand at the stranger and said, 'I hope you don't mind but I've brought a photographer with me – he's taking some pictures of me for a magazine,' and introduced

Gene Kammerman, an American photographer, who lived in Paris.

Gene was such a nice person and we all got on marvellously during the week-end – but it was, at first, quite unnerving to hear the click of the camera almost continuously as he worked on the principle that, at any second, one might get the great photograph, the unexpected expression, the unconscious but characteristic movement, the decisive moment, so he took what seemed like forty photographs a minute from underneath the piano, behind shrubs, round the corners of walls, over yew hedges and through fountains of water.

Paul, who always maintained that he loathed being photographed, was totally unconcerned and even joined in at one stage by producing his own camera and photographing the photographer. When my parents then each fetched their cameras, the four of them had a hilarious time photographing each other photographing each other!

The first day that I worked for him full-time – in May, 1960 – the telephone in his suite at the Ritz rang. He picked it up.

'Hullo?' he said. There was a pause then, 'Drop dead,' he finished. I was so surprised – and shocked – that I sat with my mouth open.

'Who on earth was *that*?' I asked.

'Some damn newspaperman,' he growled. 'They all seem to think I have twenty-four hours a day to spend doing nothing better than chatting with them. It's all due to that *Fortune* magazine article – I've had no peace ever since – and I don't see why I should talk with them, anyway.'

'But wait a minute,' I said, 'you're President of a

company that handles the public's money – it's not wholly owned by you and your family but roughly 20 per cent owned by outside investors. That means the public has a right to know who is handling their investment and what is being done with it.'

'Well, may be they can ask business questions,' he conceded grumpily, 'but that doesn't mean they can enquire into all my private affairs.'

'No, of course not,' I agreed, 'but you can't blame them if they try to. You read the newspapers – not only the business ones but the gossipy ones, too – and you enjoy them. *You* are quite happy to read about other people's private lives and thoroughly enjoy some of the scandalous bits; why should you be able to read about them but not they about you?'

He pondered a moment then his eyes twinkled. 'You've just talked yourself into being my press officer,' he grinned. 'The rules are one, keep it out if possible; two, if it's going in anyway, make sure it's accurate and three, don't bother me with it unless it's really very urgent.'

And from then on, his press coverage improved, he softened gradually in his attitude and, as many journalists, particularly in London, will know, bonds of mutual respect and appreciation formed the basis of several lasting and warm friendships between Paul and individual editors and reporters.

I suffered, of course, from comments that certain of Paul's friends made to him about my 'being eager for publicity'.

One lady friend in particular lost no opportunity to blame me for deliberately having told the press every story that was ever published about him – and about her, as well. So much so that, when she was to be involved in a certain court case, she asked Paul not to tell me as I would be sure to inform the newspapers. I am delighted to say that, without any intervention from me, several fulsome accounts appeared in the press – much, I would imagine, to her discomfiture.

In fact, my objective was to keep as many stories about Paul out of the papers as possible but I have always felt strongly that, if one asks a reporter or editor not to publish something and they do not do so, then it is only fair either to agree to their doing a story the next time they ask (provided it is accurate) or to give them an alternative 'acceptable' story.

Having said that I was to handle press enquiries, Paul certainly made sure I learnt the hardest way – by a deluge of experience.

As the news of the Jeannette Constable-Maxwell/ house-warming party leaked out to the press, they began to write, ring and call in person asking if they could come and represent their respective newspapers, magazines, radio and television stations from our country, the United States, Europe, Australia and so on and so on.

'You deal with them,' Paul told me.

'How?' I asked. 'Are they all to have invitations? Or some and, if so, how many? And what about photographers?'

'The photographs are already fixed for my own dinner party beforehand,' he said, 'but you do the rest: you can have two dozen tickets to hand out.'

It was certainly a baptism by fire! I did manage to get thirty tickets in the end but there were over two hundred and twenty press applications and it was a fearsome task trying to be fair in their allocation, particularly when, not then knowing the reporters and photographers individually, I had to double-check the authenticity of their identities. Quite a number of 'press representatives' were anything but! Some newspapers, not satisfied with only one invitation, got a dozen or more reporters to apply and newspapers in a Group would get very peeved if another paper in the Group received an invitation and they did not.

It was now that I made my first friends among the media – one or two of whom have become close personal friends of mine and have given me untold help, advice and information over the years.

I also learned that there were some of the 'other' type of reporters – not to be trusted, searching for advantages to be taken, malicious and jealous. I started my 'little black book' of such names, high among which was the American reporter at the party at Sutton Place who asked peremptorily if he could phone in his story to his London office. I later found out he made a thirty-five-minute call to the United States with a story that was inaccurate, biased and spiteful.

Having decided that I was to handle all his press enquiries for him, something of which he had had no previous experience at all, Paul had an ambivalent attitude to the whole subject.

'I don't want to be bothered,' would be followed by 'Which newspaper is he from?' and 'What does he (or she) want to talk to me about?'

As a general rule he maintained that he wanted his name kept out of the papers. I argued that, on the other hand, if it was going in anyway, it was better to go in with co-operation and an accurate story. Paul distrusted the press so much (and, to be honest, in some cases quite justifiably) that he thought they would simply publish both stories, the one we had given them, and the one they had written themselves.

However, he gradually came round to my way of thinking and, ultimately, formed some very warm friendships with particular reporters, including Robin Esser who, in the early 1960s, was the editor of the William Hickey page in the *Daily Express*; Paul watched Robin's promotion through the echelons of the *Express* with an almost paternalistic pride.

'I knew at the beginning that he was an outstanding young man of great perception,' he would say.

Another for whom he had a great regard was Richard Walter, then art correspondent of the *Evening Standard* and who later moved on to the Press Office at Christies, one of Paul's favourite salerooms.

On the other hand, there were three members of the media whom Paul disliked intensely but to whom he felt obliged to be scrupulously polite, either because he knew their superiors and did not want to provoke any embarrassing scenes or because the publicity, if properly handled, could benefit the Getty business.

His particular *bêtes noires* were certain television interviewers. There were two, one American and one English, whom he considered grossly over-familiar as, on their first meeting and in each case being much younger than he, they were calling him 'Paul' within five minutes, while he was still addressing them, he considered courteously, as 'Mr So-and-So'. He was also irked by their supreme confidence in their own importance and the fact that, as millions of viewers watched their programmes, they had the authority and complete freedom to conduct the interview in any way they wished and with questions that were frequently impertinent or rude.

When they came to Sutton Place to film, they and their staff would often commandeer the services of the secretaries, the desks, chairs and telephones. Some used to bring down their girl friends as 'secretaries'; these young women would spend the whole time smoking, asking for coffee and sitting on the desks, crossing and uncrossing their legs every time any man walked into the room and usually punctuating their own noisy conversation by asking members of Paul's staff, 'Where's Getty, then?'

Small wonder that the staff often dreaded such visits and that every now and again Paul would explode, 'Not another goddam bunch of word-maulers!'

Paul used to complain about publicity but he seemed to have a never-ending succession of people writing biographies of him, some of which never were published and, if they are now, will certainly be out-of-date. Quite often the would-be authors – professionals and 'amateurs' alike – would spend days or weeks at Sutton Place taking notes, or recording interviews with Paul, then depart, promising to send him the proofs for his approval as soon as they were ready – and that would be the last we would see or hear of them.

One American woman, in particular, to whom Paul had devoted a great deal of time giving her recorded interviews about his childhood, left Sutton Place and apparently also departed off the face of the earth. Every now and again, Paul would enquire with puzzlement whether anyone had heard any more from her, but none of us ever did.

Some journalists were thoroughly unpleasant, bloodsucking types. One television interviewer (who seemed to reckon he was the most important American alive, notwithstanding the President of the United States) arrived at Sutton Place with a pre-conceived dislike of Paul, was extremely offensive to the helpful staff, produced a really spiteful, carping interview and, finally, wrote, under another name, a very thinly disguised novel about life at Sutton Place. If the book had been worth taking seriously, it would have merited several actions for libel but Paul was advised that such a piece of rubbish did not deserve the free publicity and consequent increase in sales which litigation would have brought it.

Paul was quite pleased to have the help of ghost-writers as he said he simply did not have time, with all his business interests and responsibilities, to sit down and write articles or books himself. Apart from his professional ghost-writers, most of his personal staff, including myself, at one time or another found ourselves writing on his behalf in what we hoped would be his words. After this was done, he would very carefully go through what we had written, correcting facts where necessary or saying, 'No, that's not a word I would use.' In fact, it was always easy to spot when he was misquoted – or not genuinely quoted at all – as he had very distinctive set word-patterns and phrases and there were certain words he simply never used.

He was enormously proud of two books which he did write wholly himself – a slim volume entitled *Europe in the 18th Century* which showed his great knowledge of French furniture and furnishings of that era, and *Collector's Choice* which he wrote in conjunction with Ethel Le Vane and was a collection of some of the stories behind the acquisition of his art treasures which are now in the J. Paul Getty Museum in California.

His other writing was his diary which he carefully

75

kept up-to-date except for one occasion when we went on a business trip to France, unpacked, and found – no diary! Urgent and immediate phone calls back to Sutton Place found it still sitting in the drawer of the desk in his bedroom. It was not to be entrusted to the post, so one of Paul's top American executives from the offices in Los Angeles, who happened to be passing through London, was instructed by Paul to collect it personally and fly over to Paris with it. In the meantime, Paul was keeping his diary on odd scraps of paper, which he kept losing, so I ended up carrying them around for safety in my handbag – even when we went out in the evening!

People so often underestimated Paul because they approached him thinking they would have to assert themselves in order to get their own way; when he did not appear to argue with them, they would relax, believing they had achieved their purpose, not for one moment realizing that Paul was using them just as much for his own ends – in his own way.

A case in point was an interview he had some years ago on B.B.C. television with Richard Dimbleby. Paul had always maintained that he was prepared to suffer publicity for the sake of the company but not on his own personal behalf. (The truth of that pronouncement was debatable but it was at least what he said.) Richard Dimbleby wanted an interview with Paul Getty the man, the individual with unimaginable wealth and a trail of broken marriages who had apparently chosen to settle in this country. Paul wanted to get the business of the Getty group more widely known.

In the pre-filming discussion, the two of them reached a sort of compromise – fifty per cent of the questions would be personal, fifty per cent about business. Richard Dimbleby carefully explained to Paul that, as it was the B.B.C., no advertising could be done, no trade names (as, for instance, Getty Oil Company's 'Veedol') mentioned, and no promotion made of any of the products, processes or assets of the Group.

Paul sat gazing at Mr Dimbleby with a look of rapt, if slightly bemused, attention, nodding his head every now and again but never uttering a word – the very picture of serious concentration.

Once the programme, which was being transmitted live, started, it was obvious that the pre-arranged outline was not going to be followed. Richard Dimbleby was beginning with the personal questions and they were of sufficient complexity that a few suitably full answers would rapidly use up the allotted span of time available. Paul lapsed into monosyllabic or minisyllabic replies in order to hurry it all up. Mr Dimbleby thought this was due to camera nerves and shyness, so he talked more in order, as he thought, to put Paul at his ease.

Paul, meanwhile, was feeling that time was slipping away fast and he was not making the best use of it, so perhaps he had better be a little more forthcoming. He suddenly started to plunge into long-winded dissertations, bringing some mention of the business into practically every subject about which he was asked to speak.

Finally, at the end of the programme, when the arc-lamps were off and the technicians were gone, Richard Dimbleby remarked to me, 'Well, I think that went delightfully. He really is a charming man to interview – and so forthcoming, even about his personal life! I was warned by so many people how impossible he was to handle but I must say my nerves on that score were simply not justified,' and, with a glow of contentment surrounding him, he beamed happily as he shook both of us warmly by the hand and said goodbye.

Later, in the car, I asked Paul, 'How did you like being interviewed?'

'I thought Mr Dimbleby was very fair – on the whole,' he said. 'A little too much about me personally – but I managed to slip in a few remarks about the Company . . .'

'You certainly did!' I agreed.

'. . . and, in spite of the ban on advertising, which I just happened to forget in the excitement, I managed to mention Veedol by name three times!' he added triumphantly.

5

Humour - in Richest Vein

Paul had a very wide sense of humour; apart from his wry and dry comments – which frequently passed way over the heads of his listeners – he enjoyed simple slap-stick fun and often a mischievous gleam would appear in his eyes accompanied by a slow puckish grin. He frequently complained that he had little opportunity to play the fool as he was always expected to think great thoughts, utter oracular words and generally maintain the mien of an 'Elder Businessman'. Many people who professed to know him well would have been very surprised by the side of him they never saw.

Some months after he had moved into Sutton Place, Paul and I were sitting chatting in his study; he was in his usual, big armchair. He shifted in it, then got up and whacked the cushion hard with his hand a couple of times and sat down again.

'I like chairs to sit on that are comfortable and hard-wearing,' he said. 'That settee you're sitting on, for instance, is supposed to be strong enough to jump on.'

It was long enough to seat four hefty men side by side so it should have been proportionately strong.

'They didn't actually sell it as a trampoline, did they?' I asked.

'No,' he said, 'but they said it was strong so let's try it.'

He got up, took off his shoes and climbed up beside where I was sitting. 'Come on,' he said, pulling me up, 'you try, too.'

I protested that two adults jumping up and down were not going to do the springs any good. 'What if they give way?' I asked.

He grinned. 'I'll send it back.'

So for a good five minutes, we bounced up and down – or rather, he bounced and I nearly killed myself laughing as he did an Indian war-whoop each time he went up in the air.

Suddenly there was a knock at the door which, luckily, was closed. Quick as lightning, we were off, into our shoes, he in his chair, me standing in front of the fireplace, both of us trying not to pant too hard and to stop giggling.

'Come in,' called Paul.

In came a couple of his business associates, they a little deferential to him. Paul gravely shook their hands.

'Sit down,' he said. 'The settee's very comfortable. I recommend it. It's strong, too.' And he gave me a big wink in an otherwise deadpan face.

One of Paul's rarely known talents was mimicry, both of 'types' and particular individuals. With all but his closest friends, he used to say, he felt ill-at-ease and unable to relax, always aware of how he was expected to behave and that he was not expected to make a fool of either himself or other people, so he could only be persuaded to perform his imitations, which were sometimes cruelly accurate and at other times extremely funny, to a very few people, not all of whom really appreciated them.

80

One of his best acts was a devastating version of Hitler delivering a speech at a youth rally. Not only could he reproduce the voice, the intonation and the rise and fall in pitch from normal tones to the almost falsetto scream, but he would pull his face about until he almost looked like his subject and strut and gesticulate with all the movements only too well known to some of us. He was so good that his imitation really ceased to be funny at all.

Another of his voices was what I suppose might be called very upper-class English county; it was based on several of his acquaintances but one neighbour in particular. It was that intonation that at times can be so infuriating in its apparently exaggerated la-di-dah vowels, slightly nasal timbre and which sounds like speaking through a mouthful of pebbles; the result is that one can hardly understand a word. Paul's rendering was without trace of American accent and accompanied by the correct slightly disdainful look and limp handshake.

He had quite a wide variety of American accents which varied from the slow southern drawl to the 'de and dose' of Brooklynese, not forgetting the top echelons of Boston, the formidable American aristocracy of the social register. This last he would deliver as beginning with an enquiry why one was not wearing white gloves, continue with a lecture on how to lay a table with cutlery and wine glasses for a fourteen-course meal and finish with a discussion of just who could or could not be invited and the reasons for the discussion, which would leave one wondering whether or not he was still exaggerating!

He also did some pithy scenarios of mock conversations he had with other businessmen or some of his own executives, most of whom he considered did not have enough sense to 'come in out of the rain'.

He used to make me roar with laughter at his version of my having an argument with him. For instance, he would say that if I were disagreeing with him, I would throw in one reason after another, usually prefaced by the expression '... and another thing...' His examples were so good that I used to finish up by remarking that, if he knew my reasoning so well and could deliver it so plausibly, why did we have an argument at all when it seemed he was already converted to my way of thinking?

'Why', he would answer, 'I would miss the chance of some mental stimulation. If I seem to disagree with you, you give me so many reasons why I'm wrong that I know, if I later have a similar discussion with someone else, that I only have to remember your views and use your arguments and I'm not going to get cornered!'

Paul was rather chameleon-like: he changed according to the company he was with which, I am sure, is why my parents and I saw so much of his humorous side. We have each always had our own types of humour, my mother having the shrewd, quick Scottish wit and art of mimicry, my father being a widely-known raconteur and after-dinner speaker and I, also, a mimic. I have several accents which I can reproduce, particularly Indian, thanks to tuition from some Indian friends of mine. Indians have such a marvellous sense of humour and are one of the few peoples I know, apart from the Scots and English, who can laugh at themselves.

Paul himself had many Indian friends, among whom was one lady whom he liked and admired for her charitable works but he usually tried to avoid

82

her as she was extremely talkative. If we met her at some social function, he would draw me into the conversation and then disappear himself. I liked the lady very much and thoroughly enjoyed talking to her but Paul's evasiveness embarrassed me, as I was worried in case she noticed and her feelings became hurt.

One day, after we had had one of these encounters and his departure had been even more hasty than usual, I thought I would teach him a lesson. When he returned to Sutton Place, I telephoned him from London, and persuaded the girl on the switchboard to put me straight through to him, without announcing my name.

'Oh, Paul, dear,' I began in my best Bombay accent – and went on and on for nearly twenty minutes, having the conversation that Paul ought to have had earlier in the day. Eventually, I had to start chuckling at his great politeness, particularly when I knew he was thinking the conversation was likely to go on for another half-hour or so – so I dropped my accent and owned up.

At first, he couldn't believe it had been me, then he started to laugh.

'Well,' he said, 'as you succeeded in fooling me for twenty minutes, I guess I shall have to pay a penalty! What shall it be?'

'You can be kind and ring up your Indian friend to atone,' I answered, 'and you can always explain, if the conversation becomes too lengthy, that you have to make another business call, which is always true enough, anyway.'

An hour or so later, he phoned me back. 'I've done my penance,' he said, very cheerfully, 'and, as it happened, it wasn't nearly as bad as my conversation with you – it only lasted ten minutes and she was very sweet and charming.'

'There you are then,' I said smugly, 'and in future, you can spare a little longer to talk to her!'

'I shall,' he replied 'but I am also going to get my own back on you, for hoodwinking me for so long.'

After that, he often used to telephone in some strange accent and I had to listen very carefully in order not to get taken in myself, or make the mistake he did.

Some time after my phone call to him, his Indian friend rang him – and for the first few minutes he thought it was me! Luckily, she was so good-natured that she instantly accepted his explanation that, due to deafness, he had misheard her name when it was announced by the girl on the telephone and thought it was someone else!

'Now I'm terrified every time I answer the phone,' he said, 'wondering whether the lady the other end is you.'

'Never mind,' I answered, 'it gives you that mental stimulation you're always saying you want.'

When Paul visited us in Kent, we would retire after lunch into the comfortable chairs in our well-sheltered little garden and relax and chat. Away from the formalities of London social life and restraining influences of Sutton Place, Paul would forget about keeping up appearances and his mournful, 'buttoned-up' expression would disappear. Instead, his eyes would begin to sparkle, a small smile would widen to a lopsided grin and he would join in the family teasing and banter, giving as good as he got. Before long, he would be playing the fool with an uninhibited gaiety that would have dumbfounded some of his friends who thought they knew him so well.

Between them, Paul and my father perfected a splendid rendering of a couple of drunks who could occasionally be seen lurching around the garden of one of our local restaurants, sagging against an old lamp-post on the terrace or carrying on a very serious conversation with plenty of hiccups amongst the slurred speech.

One day, Paul decided that the garden of one particular restaurant needed the finishing touch of a piece of statuary.

'I think a nice life-size figure in stone,' Paul said, 'somewhere about here,' pointing to a large area of lawn. 'I'll take the part of the statue and give you an idea of how it would look.'

He proceeded to take up various positions in different parts of the garden with wobbly arabesques and modernistic 'pose with chair in air over head' or a type of seated Buddha on the lawn. My mother managed, between laughing, to take one or two photographs amid all the fooling around and we still find them funny now.

Paul was very fond of statues – particularly of the Herculaneum classical era, and he had many such exhibits in his museum in Los Angeles. They were much too valuable to be continually moved around or kept out-of-doors in our unpredictable climate. So when the 5th Duke of Sutherland, as had been agreed at the time of the sale of Sutton Place, removed a rather bloodthirsty, large stone statue at the southern end of the yew walk off the South lawn, and left only a broken-up circle of stone blocks which had supported the plinth and statue, Paul pondered long and frequently what to buy to put in its place.

He couldn't make up his mind how tall the replacement should be. Too short and it would be out of proportion with the long yew walk and would therefore not draw the eye; too high and it would, as the original statue had done, obscure some of the lovely, extensive view over the wild garden, which lay below along the deeply inclined banks that run down to the river, and beyond over meadows to Guildford and the distant hills beyond.

Eventually, Paul more or less decided he would have a lower group of figures of sufficient breadth to appear substantial and well-balanced but not high enough to hide the distant scenery.

He was explaining to my parents what he had in mind. 'Come on,' he suddenly said, 'I'll show you,' and he grabbed me, placed me firmly in the middle of the stone circle, directed my position like a film producer – most of the time I was balanced on one foot – and then proceeded to cavort around me taking up various contortionist positions which, he claimed, were simply copies of Henry Moore or other modern sculptors and who, as Paul frequently remarked, seemed to have as models either some unfortunately misshapen people or some peculiarly agile yoga addicts.

As it happened, Paul never did replace the statue. He was always going to get around to it 'tomorrow' or 'next week' but, eventually, I think he simply preferred the peace of the uninterrupted view down the long green lawn to the tops of the multi-coloured flowering shrubs and trees in the wild garden.

Paul and I shared a great love of swimming and the two pools at Sutton Place were in frequent use all the year round. It was one of those scorching hot days with an intensely blue sky when Paul decided to play truant from work and go swimming and, so that he would not be disturbed with telephone messages or other unwanted interruptions, we sneaked out of a side door and round to the out-door pool.

After each swimming a few lengths, we started, as always, fooling around. I was looking up, admiring the sky, when he pushed me under the water; I came up spluttering and caught his ankles as he was trying to swim out of reach and pulled him backwards – so he splashed me and I pushed him under.

At this stage, his friend Mary Teissier walked into the garden, looked appalled and rushed to the side of the pool.

'Robina, Robina,' she screamed, 'you mustn't do this – you are too rough and he is an old man.'

Paul surfaced in time to hear her last words and, treading water warily with his back to her, muttered, 'She makes me feel even older than I am – she'll only be happy when I'm in a wheelchair wrapped round with rugs.' Then he grinned, took a deep breath, gave me a watery-eyed, protracted wink and slowly disappeared beneath the water again.

Mary, of course, as he had intended she should, thought I had pushed him under again and became very vituperative in her comments. As he was by now tickling my feet under water, all I could do was shriek with laughter – and also disappear under the water in a cloud of bubbles.

Paul, when he surfaced, started to swim to one side of the pool as Mary rushed round with his towel, begging him to come out. I am sure it was partly water in his ears but he was apparently a little deaf that day and didn't seem to hear her – or see her, for that matter. He changed his mind and swam towards the other side of the pool so, naturally, Mary hastened round there. He changed his mind about whether to come out of the pool and, if so, at which side, several times during the next few minutes and Mary got quite tired running round the pool, calling to him and clutching his towel.

Eventually, he climbed out. He said afterwards, 'There are some people who just won't let me do what I want and fool around if I feel like it, and I can't bear screeching women.'

Paul and my mother's mother, who was several years older than he, got on very well in spite of each being a little deaf; Paul was always very charming, very kind and full of concern for, and interested in, people older than himself and, while she was alive, my grandmother was always included in our family visits to Sutton Place. He seemed to have a wistful envy of the closeness of our small family, frequently remarking that, in spite of so many children, and grandchildren, he felt he lived more of a bachelor existence than anyone else.

As an accomplished exponent of the wry or dry comment, he enjoyed my grandmother's pithy comments, delivered in the Scottish accent that she had never lost during fifty years of living in England, and would respond in the same vein.

The five of us were at dinner one evening in Rye, soon after his arrival in England; my grandmother, who had heard a great deal about him, had only just met him but, like all Scots, she was quick to sum him up and appreciate his sense of humour.

'Mr Getty,' she said, suddenly, 'did you know Robina can bark?'

Without blinking an eyelash or twitching a face muscle, he replied, equally gravely, 'Why no, Mrs Audsley, I guess I didn't, but I shall look forward very much indeed to hearing her.'

And turning to me, he continued, 'I must be a little out-of-date about the accomplishments of modern young ladies; I don't think they learnt how to bark, moo or neigh in my teenage days, though I know several who can bleat on occasions.'

The next day, when we were back in the privacy of our country garden, he demanded a demonstration of the barks of the varying breeds of dogs, which I could imitate well enough to get an unfailing response from other dogs and to chase away cats.

When, some months later, he had bought Sutton Place and the first five Alsatians as guard dogs, we went round to see them in their kennels. One started barking and another howling.

Paul grabbed my arm. 'Imitate that,' he ordered, a glint of devilment in his eyes.

I did and, in a very short time, all the dogs were barking furiously.

'That's very good,' he said. 'I think you should practise your Alsatian barks, they might come in very useful; if an intruder thinks you've got a large, fierce dog with you, he's more likely to go away. Also,' he added looking at me sideways 'if you bark, why do I need to keep the dogs? Think what it would save me in kennelling and food!'

Sometimes, when we were discussing some social topic or a question of manners or behaviour, he would suddenly break off and, viewing me over the top of his spectacles, remark, 'You should feel honoured, dear, that I'm prepared to have you English know me. It's a sign of my truly noble character that I'm prepared to mix in all company, even yours.'

'Ah!' I would retort, 'but are *we* prepared to mix with *you*?'

At this stage I would get bombarded by him with screwed-up bits of paper – promptly thrown back by me until, twenty minutes or so later, breathless and giggling in a thoroughly adolescent manner, we would call 'pax' and, while he went and got us a drink each for our thirsty throats, I would do my best to clear up the mess. Mind you, he never *was* very tidy anyway.

Paul was immensely proud of his son Gordon's ability as a pianist and composer and, although he was understandably disappointed when Gordon left the company, he hoped that he would pursue his musical career and was very sad when Gordon did not do so. However, there were for a while in the Long Gallery at Sutton Place, two beautiful grand pianos which belonged to Harold Freese-Pennefather who was, in the early 1960s, the British Ambassador Extraordinary and Plenipotentiary at Luxembourg. Mr Freese-Pennefather had been unable to take his pianos with him and Paul was only too pleased to look after them.

Paul several times expressed his regret that, as far as the Arts were concerned, he was strictly an onlooker and completely incompetent as a practitioner of any sort. In fact, with a little more application in his younger days, he might have made quite a passable pianist. As it was, lack of practice and a failing memory for the music had produced one of his best party turns – which he only performed for a few close friends.

He would sit at the piano and, with vastly exaggerated care, adjust the height of the piano stool, sometimes even getting up to examine it from a distance or to produce an imaginary tape measure. Having at length apparently satisfied himself, he would then sit down, draw up the stool to the keyboard and remove his jacket, watch and cuff-links and roll his sleeves up to the elbows. After flexing his fingers and shaking loose-wristed hands in the air, he would smile amiably at his audience; suddenly, this would be replaced by a frown and, biting his lip, he would get up, walk round to the side of the piano, peer in under the strings, take a very deep breath and blow – hard – at some invisible dust inside. Reassured, he would return to the piano stool, place his feet on the pedals, his hands on the keys and with strong, unwavering fingers, begin.

Fortissimo, a succession of cathedral-like chords would surge out with a deep resonance – for about three bars. Then, hands still in mid air, he would look round slyly and say, 'That's all I can remember,' and would shut the piano, replace his watch, cuff-links and jacket and regard with amusement the nonplussed expressions of those who had not seen his performance before.

The fooling around nevertheless was a cover-up for his genuine envy of people who were, as he put it, 'practitioners and not ponderers' of the Arts. As, long before I qualified as a lawyer, I had been trained as a concert pianist, I still remembered much of what I had learnt, in spite of lacking the time for any regular practice, and Paul would often propel me towards a piano, partiularly after dinner in the evening, and sit and listen to whatever I might play for as long as I went on, half-an-hour or an hour, sometimes even two hours.

When Gordon came to stay, Paul would look forward all day to hearing him play either his own very appealing and intriguing compositions or the classical music that Gordon and I both played. Paul used to enjoy having a recital from both of us, each playing a piece alternately; sometimes Gordon and I would simultaneously play a Schubert impromptu which we both knew, on the two pianos; I gather it sounded quite impressive as, if one of us missed a note, the other usually played it.

When we were having one of these evenings, Paul would shepherd any house-guests up to the Long Gallery to listen, regardless of whether or not they might want to. Penelope Kitson used to get totally bored and would often sigh, 'Oh, not again,' and try to find some excuse or something that needed to be done urgently and which, such a pity, would prevent her from coming up to hear the music that evening – but, next time, she would love to.

I was probably one of the very few women, if not the only one, to whom Paul proposed marriage regularly each year. This was another of our shared jokes – and one which I do not believe he would have dared to try on most of the ladies he knew!

Paul had acquired from somewhere his own ritual for St Valentine's day: he maintained that, if the woman asked the man to marry her and he refused, he had to give her a silk dress and if he asked her to marry him and she refused, she had to give him a pair of gloves.

The first time he told me this, I didn't really pay much attention. 'I knew the silk dress part,' I said, 'but I've never heard of the "forfeit" of the gloves.'

'Oh, yes,' he insisted, that was the way he had

J. Paul Getty and Sir Thomas Lund relaxing at the local inn

Long view of the Lunds in Kent

Short view of J. Paul Getty and Robina Lund at Sutton Place

always been brought up to do it.

I never gave it another thought until suddenly, on the next 14th February, when I walked into his study with some papers, he suddenly fell on one knee and, pressing his hands together and with rolling eyes and a lugubrious voice, said, 'Will you marry me, Robina?'

I simply could not keep a straight face as he started to pull funny faces, waggle his ears, and give me a gap-toothed leer.

'Certainly not! You're too young for me,' I eventually managed to get out between my fits of laughter.

The words were hardly out of my mouth when up he stood, put on his spectacles and fixing me with a penetrating gaze, solemnly intoned, 'Right then, dear, you owe me a pair of gloves. Ah-ha!,' he added gleefully, 'I caught you out! I asked you to marry me and you refused, so I want my gloves.'

'All right,' I agreed, 'I fell right into that trap, I admit. But gloves? You never wear them.'

'I might do – but I still want my gloves anyway,' he insisted.

So I gave him his gloves – I did not have any peace until I handed them over. Until then, it had been, 'You won't forget my gloves, will you?' every two or three hours during the day for two whole days.

I bided my time for 'my revenge'. It was a tacit agreement between us – and a basic part of the fun – that one good trick played by one of us on the other deserved a return match.

The next year, on 14th February, I waited in the morning until I thought he would be awake and telephoned him on his private line. I mimicked the voice of a lady-friend of his who was very keen on him but whose feelings were definitely not reciprocated; in fact, he went to a great deal of trouble to avoid her and was for ever being diplomatically 'out', 'away', 'engaged' or 'travelling'.

'Paul, darling,' I breathed down the telephone, 'I know you are such a sensitive person, you are so shy, you will not say it yourself . . .' and so on and so on, until I got to 'so now I ask you, my darling Paul, knowing you feel as I do, will you marry me and always I will look after you, never leaving your side?'

There was a stunned silence at the other end of the telephone and it was all I could do not to burst out laughing. Eventually, after some throat-clearing, Paul spoke, 'Uh – well, dear, I – uh – appreciate your very generous feelings in wanting to look after an old man but – uh – I – uh – I feel that I'm not too good a bet with five failed marriages behind me so, dear, no, I won't marry you, it wouldn't be fair to you.'

'Oh fine,' I said in my normal voice, 'that's one silk dress you owe me.'

'Robina! I should have guessed,' he exclaimed. 'Just wait till I see you – and now I'm never going to know if it's you or her on the phone, you wicked girl!'

I was allowed to choose a dress, so he did keep to his own rules of the game.

The following year, I decided to telephone him at one minute past midnight with my proposal but his number was engaged. I must have tried six times during the next half hour – always the engaged signal. Eventually, my telephone rang and it was Paul.

'I've been trying to get you for half-an-hour,' he complained.

'But *I've* been trying to get you,' I retorted. We both had had precisely the same idea so we agreed to exchange presents. I was a bit bored with giving gloves and he had no objection to substituting another 'forfeit' for the silk dress so we each gave the other a small toy animal – a lion from me to him, a dog from him to me. After that, the 14th February became an annual ritual like another birthday and Christmas combined and since Paul, once he had built up a tradition, was very superstitious about continuing to observe it, we always went on exchanging Valentine gifts after that.

6

Travels and Travailles

Paul was very unreliable in his travelling habits – a characteristic to which we all had to adapt. It was quite simple for him to decide to move on at short notice as he usually had only two cases to pack but I was always in charge of the business papers when I travelled with him and, consequently, had not only my own luggage but multitudinous letters, files and reports with which to cope.

During one of our European business trips Paul had decided to leave Munich for Vienna at nine o'clock one morning. The previous night we were at a birthday party and we left very late at half past three in the morning. (We later heard that the party didn't break up until eleven o'lock later that morning!) It seemed to me that, at this hour, there was little point in going to bed as I would need a minimum of three hours to pack everything up and another hour or so to bath, dress and have breakfast. So I stayed up, got everything done, had my breakfast and sat and read the paper; and sat; and sat. Eventually at half past ten, I telephoned through to Paul's room. No answer. I went along and knocked on his door. Still no answer. Now beginning to worry, I confirmed that the car was still in the garage; once again, I got the hotel operator to ring his room. At last, with relief, I heard the telephone picked up and an incredibly low, hoarse voice foggily mutter, 'Hullo?'

'Paul,' I said, 'you are supposed to be on the way to Vienna and you wanted to leave at nine this morning.'

'Well, what time is it now?' the voice croaked.

'Nearly eleven o'clock.'

'Oh my! You'll just have to postpone the meetings in Vienna. I'm too tired to leave today. I'll see you downstairs at one.'

So I spent the next two hours re-arranging the business schedule in Vienna, which was quite tricky as I had no idea when we would actually arrive there.

At one o'clock, I was sitting in the restaurant; at half past one I felt I might as well order my meal as I was by then so hungry; an hour later, just as I was having my coffee, Paul appeared, having woken up again only half-an-hour before. The result was that *he* was fine for the rest of the day but, by five o'clock, *I* had to retreat in order to get two hours' sleep.

After that, I never again packed in advance but instead always waited until Paul started to pack his own things. I soon learned that his usual hour of departure was very rarely before half past ten in the morning.

Paul, having his own peculiar method of travelling, found it hard, if not impossible, to adapt to anyone else's. What was particularly irritating was his habit of not listening, in the sense of paying attention, to what one said, once he had formed his own pre-conceived idea.

A typical example was the occasion when I was to join him in Baden-Baden. As, like him, I did not fly, I was travelling from London by train and boat and having a very precise schedule of times of arrivals and departures and changes of train, I cabled Paul that I

97

would arrive at Baden-Baden station on the Thursday morning. I did, with all four trunks and cases of his papers as well as my own belongings but – no Paul, no car. After everyone else had left the station, there was just one porter and one taxi left, so I shrugged my shoulders and agreed with the porter that there did not seem any point in waiting around longer and he loaded the luggage on to the taxi.

When I arrived at the Brenners Park Hotel, I really was surprised to find that they were expecting me; in fact, my arrival caused considerable merriment as they said that Herr Getty had left ten minutes earlier for the station. Obviously, he would return to the hotel sooner or later and there was no sense in chasing after him. I was pretty sure he would not think of telephoning back to see if I had arrived as that sort of thing never crossed his mind – so I thought I might as well have the bath to which I had been so looking forward after the long journey.

I was just in the middle of luxuriating in the lovely warm water when there was a thundering at the door and a peeved voice called, 'Are you there, Robina?'

'Yes, Paul,' I yelled back, 'but I'm in the bath – I'll see you in half an hour.'

When we met later for coffee and explanations, he told me a typically complicated story of why he had not met me: the doctor had come late to see his vaccination, then he had rushed down to enquire from the Head Porter at which station – Baden-Oos or Baden-Baden – the train would arrive and been told that the connections from London would come in via Frankfurt and arrive at Baden-Oos.

'But, Paul,' I expostulated, 'I cabled you specifically

that I was coming via Basle and arriving at Baden-Baden. Didn't you get my cable?'

'Well, yes,' said he, doubtfully, 'but I didn't think you necessarily meant to arrive at Baden-Baden – most people get confused by the two stations and get off at Baden-Oos.'

'In future,' I said firmly, 'please assume that I am going to be where I say I am and that I am going to do what I say – if then I don't, that will be the time for you to get annoyed.'

I must admit that he never made the same mistake with me again and, from then on, he was almost always on time – for me at least.

Our journey to Munich from Baden-Baden was the first time I had driven a long distance on a motorway with Paul and it proved to be a memorable trip. Paul was driving the 1955 Cadillac and I, although on my first visit to Germany, was acting as navigator. In those days, in 1960, Paul was a fast driver and we travelled most at between seventy and eighty m.p.h. At one stage we passed what we estimated must have been a ten-mile-long military convoy: armoured trucks, Red Cross vans, troop carriers, out-riders and many more, all travelling in strict formation in the slow lane with their headlights on. It took us nearly half-an-hour to overtake the column. We were thankful when we had passed them.

After we had been driving for about two hours I turned to speak to Paul and noticed the fuel gauge was registering 'Empty'.

'Is the petrol gauge working?' I asked.

'Yes, dear, why?'

He glanced down and his face froze in disbelief and astonishment. There was great panic (and considerable scorn about Cadillacs from me) when I discovered he had no reserve supply. Paul thought there were probably two gallons left once it registered 'Empty' – the only point being that we did not know for how long the indicator had been showing 'Empty'.

'You'd better check in the owner's manual, dear.'

So I opened the glove compartment and out fell Paul's usual clutter – driving gloves, glasses, dusters, a spare pair of socks, a tie he had forgotten to pack, two giant bars of Cadbury's nut chocolate, two apples, a broken electric shaver, a camera and the usual assortment of maps, museum catalogues and tourist guides. I found the manual at last and tried to pack the rest back. The manual confirmed there was no reserve switch, and that the most economical speed would give us about twelve miles to the gallon. So, with luck, if the gauge was right, we might have twenty miles in hand.

After about ten miles we saw an Esso station – but the other side of the road and, of course, crossing the central reservation was absolutely forbidden. A few miles farther on, on our side, was a United States Army filling station: I was all for pulling in but Paul simply would not dare.

So on we drove until in desperation (and after some chivvying from me), Paul turned off at the first side-road with a sign-post purporting to direct us to a village 2 kilometres away. It was the first sign of any village for miles, but was not mentioned on our maps. It was after a very long 2 kilometres that we came to a fork without sign-posts; Paul decided to take the right fork. After a mile or so, the smooth surface changed abruptly to gravel and in no time we found ourselves on a mud track in the middle of a large field surrounded by fir trees and with not a human being or animal in sight.

We managed to turn back and took the original left fork, which began as a rather muddy track through forest but led on to a narrow, but well-surfaced road – and, ahead, the 'village'.

It was beautiful – six houses one side of the road, two and a pond the other, the houses all in pale pink and grey stone set in green grass and overhead a pale blue sky; there was a sweet earthy smell and there was absolute peace and quiet.

But no petrol pump – the nearest one was at another village four miles on! Once more we set off. The road now was good and fir trees alternated with harvested fields and green meadows on each side of us. We had gone about two miles when the engine spluttered, jerked – and went silent. We coasted to a standstill.

'I guess I'd better walk to the village and bring a can of petrol back,' said Paul. 'Are you coming, dear?'

I viewed the road, which ran straight for about a mile then curved right and out of sight behind a mass of trees, thought of my high heels, heavy coat and all our valuables in the luggage.

'No,' I said, 'I'll lock myself in the car and stay with the luggage until you get back.'

'Good, dear. I'll try and find someone to drive me back.' I watched him disappear and sat back to wait. The air was soft and fresh, a pale sun sparkled through the trees and there were rustlings and crackles in the undergrowth that I assumed (and hoped) were squirrels and hares. In the far distance I could hear the comforting chug-chug of a tractor – at least someone was within earshot of the car-horn.

It was only about twenty minutes later that a small incredibly muddy Volkswagen appeared ahead, came tearing up, braked hard, and stopped nose to nose with the Cadillac. Out climbed Paul, still immaculate in his dark blue overcoat, together with a scruffy, oily mechanic with a 10-litre can of petrol. Once this was in the car, we followed the Volkswagen into the village and filled up properly. I think that, for the first time in his life, Paul really appreciated the value of petrol and we agreed that, in future, *I* would watch the petrol gauge.

The mechanic directed us on to the autobahn which we joined after only ten minutes' drive – just fifty yards behind the last vehicle of the army convoy!

Although he was a very nervous passenger in a car and would frequently climb in with an immediate admonition of 'not too fast' to the driver, Paul was, as I have already said, a very fast driver on occasions. Once we got on a good autoroute in France or the autobahn in Germany, he would put his foot on the accelerator of the Cadillac until it was doing up to 100 m.p.h. Normally, he drove quite well; being a cautious driver inclined, if anything, to brake nervously and unnecessarily. At speed on a clear road, his nervousness seemed to evaporate proportionately with the increase in speed, and I got more on edge particularly as he *would* sit with his right hand on his knee, his left elbow resting comfortably on the arm-rest of the door, holding the steering-wheel lightly at the base with three or four fingers.

Being power steering it was, of course, very light and I used to ask him, 'At this speed, what happens to us and the car if you suddenly start sneezing?'

When I had pointed out to him that he made me as nervous as other drivers made him, he agreed to keep both hands on the steering-wheel, except when he was eating an apple or chocolate (Heaven forbid that we should ever stop in a lay-by for a snack!) when he would drop speed to about 50 m.p.h.

Paul driving in a strange city was quite hair-raising in a totally different way. A typical example was our entry into Munich. It was my first visit there so I obviously was no help at all in navigating. Paul had not driven there himself for more years than he could remember and we had only main route maps and no local street plans.

At first, driving through the western suburbs, all went quite smoothly as we followed the wide, well-surfaced gently curving road. The nearer we got to the centre of Munich, however, the busier the traffic became, and all the drivers seemed to drive very black, very shiny cars, very fast and very close together. Every time one overtook us with about one foot clearance, Paul would instinctively jerk the steering-wheel and we would swerve towards the kerb. The trouble was that he was so busy looking for traffic directions that he never looked in his rear view mirrors and had no idea of what was coming up from behind – and he didn't care for side-seat driving unless he asked specifically for navigational help.

Eventually, we came to what was obviously a major road junction with plenty of traffic in all directions and traffic lights green in our favour. We were just reaching the centre of the junction, driving very slowly, when he stopped altogether to see if he could make out the names of the streets. The cars behind hooted and one or two drivers, who managed to squeeze round us, hurled what I imagined to be some fairly pithy comments at Paul – at any rate, they shook their fists and arms energetically.

103

Then the lights changed and the cross traffic started off with a cacophonous roar as though it were in a Grand Prix – and we were still sitting stationary in the centre of the road. The only comfort I had was that we were in a very large, heavy, cream-coloured car and the chances were that no one would miss seeing us, (although I was not quite so confident that we would not miss being hit), and if they did, I decided we were heavy enough to absorb quite a hefty bump without suffering much damage to ourselves.

As it happened, our size was our saving: we blocked so much of the road that soon no one could move so, while the traffic waited and to the accompaniment of car horns, tram bells, shouting drivers and pedestrians – and me imploring him to move in any direction, just so long as he *moved* – he, in slow and stately fashion, quite unaware of the surrounding noise because his total concentration was pin-pointed on the one object of where to go, turned the car to the right.

It turned out to be the correct road and, once satisfied of this, he sat back and smugly said, 'Well, there you are, what was all that fuss about? This is the road I was looking for.'

'It was just,' I replied, 'that we were holding up rush-hour traffic at a major road junction when the lights had turned against us and the drivers were getting very angry!'

'Oh really?' he said, 'I didn't notice.'

There followed the collapse of one nervously exhausted passenger!

I sometimes took over driving from him, apart from when we entered a strange town, at which stage I was instructed to change seats and return to acting as navigator. Curiously enough, although he frequently pored over maps and plans, Paul was not very good at navigating. It was as though he were unable to relate the scaled-down drawing to the architectural realities.

He was usually completely relaxed when I drove, although he would sometimes get into a panic that we were on the wrong road and insist loudly, 'Stop! Stop here! Pull in there, no there, beyond the next car,' and so on.

The secret was to maintain a total calm and ignore all his directions or else one could find oneself in no end of sticky situations.

The first time I drove him, in my Citroen I.D. 19, he gave his usual instruction of 'not too fast', so we started off smoothly and decorously up to 20 m.p.h. After ten minutes or so (and this on a dual carriage-way), I edged it up to 30 then 40 m.p.h. He started inspecting the dashboard controls and asking questions.

'If you want me to show them to you properly, I'll pull in at the side and stop when I can,' I said, 'but I can't look at the dashboard, look through the windscreen and look in the mirrors at the same time.'

'No, no, quite right, dear – you just keep on driving now, we're making quite good time.'

After a few more minutes, we were doing 45 m.p.h. and he suddenly remarked, 'I think you could go a little faster now.'

'How fast do you want to go?' I enquired.

'Well, I don't mind if we go up to thirty-five or forty as you're a good careful driver.'

'How fast do you think we *are* going?'

'Oh – about twenty-five or thirty?' he replied with mild interest.

'If you look out of the window,' I said, 'you will see

that we are doing rather more than that. In fact, we're now nearly up to fifty.'

'No!' he exclaimed. 'I can hardly believe it. Where's the speedometer?'

(Although he had only just been inspecting the dashboard, he was extraordinarily unobservant about these details, except on his own car.)

'Almost in front of you,' I said.

He peered down at it shaking his head and then looked out of the window as we overtook a couple of vehicles.

'I certainly have no feeling of speed,' he observed, 'I'm very comfortable.'

And he stretched out his legs, put his head on the back of his seat and in five minutes was fast asleep. He never again worried about my speed and would often sit reading out loud from newspapers and books while I drove.

The drawback of making a trip with Paul in his car was that one was totally dependent on him and his changes of mood, as my mother and I learned to our vexation when we took him with us on holiday one year.

He travelled as he behaved at museums – in that he would do precisely what he wanted regardless of other people's preferences or arrangements. Rather than adapt himself, he would leave.

Now, as a family, we were very fond of staying at a delightful country house-type hotel at Castle Combe in Wiltshire. We had been on holiday there once when Paul was up in London and he became quite intrigued by the sound of it as he heard what we had done during the day when he telephoned each evening. So when, some months later, my mother and I decided, while my father was abroad on one of his frequent business trips, to have ten days' holiday there, Paul eagerly asked if he could come with us.

'We can go in my car,' he said. 'Lee (the chauffeur) will drive us and that means he can have a change of scene, too.'

With some difficulty, as the hotel was not large and thus nearly always fully booked, we succeeded in getting him a room and off we set.

The next morning, Paul had his inevitable business telephone calls to and from the hotel but we eventually managed to get him out for a walk down the very attractive main street with its soft pinkish grey stone-built houses on each side. We stopped for coffee and home-made flapjacks in the little coffee-house that my mother and I knew well from our previous visits; Paul enjoyed his flapjack so much that he had four!

In total contrast, by the afternoon, Paul had had one of his mercurial changes of mood; the morning's easy humour and pottering about had given way to impatience and restlessness. He did not want to have any more telephone calls, nor stay in the hotel, nor go for a walk (all of which he had himself planned in the morning) but he had to get in the car and keep on the move – anywhere, it did not matter, as long as we were moving – and he did not want to go on his own; we all had to go.

So it was that my mother suggested we went to visit Corsham Hall, which is a beautiful old house but in those days of 1964 it really had only one public room with paintings and furniture on display. Paul walked in, round, out and stood by the car; he had seen what he wanted to see and he was champing at the bit to be on the move again. My mother and I, who were also interested in the architecture and the grounds (to neither of which did he give more than a glance), firmly continued our tour as a matter of principle but it does take away from the enjoyment when one's companion is pacing up and down looking at his watch every minute or so.

By the following day he had completely changed temper yet again and we took him to Berkeley Castle, which appealed greatly to him as it was large, impressive and old. He was very keen for me to take multitudinous pictures of him with the castle as a background which I did – until it was my turn, after an hour or so, to become bored and uninterested – and he was quite upset when I suggested we went to look at the terraced gardens and grounds instead.

That evening, he was back on the telephone at the hotel and announced to us that he would have to leave the next day and return to Sutton Place because an elderly American whom he had not seen for years and whom he was unlikely to see again was about to arrive unexpectedly.

My mother was astounded by what to her Scottish upbringing was a display of bad manners on Paul's part – particularly bearing in mind that he had invited himself and that he knew that the owners of the hotel had had great difficulty getting him in. Also the American had suggested with enthusiasm that he should come down to see Paul in Wiltshire.

However, all she did was to point out that if Paul went back, we would have no car – and it was a place where a vehicle was absolutely essential, being a village built in a bowl-shaped valley, the only road in and out of it in each case going down a long and steep hill and, on the far side, up an equally long and steep hill.

'I'll send the car back for you,' promised Paul.

However, we had had too many examples of his unreliability and could see ourselves marooned there indefinitely, so my mother, with some annoyance, decided we would all have to go back together the next day, which we did.

'But it's the last time,' she said, 'that we ever go on a journey with Paul without having our own car.'

And it was.

As a tail-piece to this episode, it was only three weeks later that Paul suggested we all three left the following day to continue our interrupted holiday and was amazed when we explained that the hotel was now fully booked and, in any case, my mother now had commitments to keep her in London. He was so used to taking instantaneous and idiosyncratic decisions to pack his bags and move that he could not comprehend that other people had more ordered existences.

Sometimes his abrupt changes of mind did have a reason – of sorts – behind them.

Paul had a pathological terror of being near anyone with any infectious illness, however mild it might be. Whenever a member of the staff caught a cold, she or he was either kept well out of his way, or else had to muffle it as much as possible, avoiding any sneezes, coughs or grunts. On the other hand, when Paul caught a cold or 'flu and himself had a temperature (to which he always referred as a 'fever', causing certain consternation to the uninitiated, who pictured him tossing and turning with a temperature of 104° or more), he made no effort to keep away from anyone else and, in fact, would become quite sulkily morose if one avoided him.

Once, when we were in Paris, I developed a mild

attack of 'flu. I woke up one Sunday morning with the usual aching bones and crashing headache but my temperature was only 101°, so I rang through to Paul at about ten o'clock, by which time I thought he would be awake, and said I had a touch of 'flu, probably the twenty-four-hour type, and thought I would spend the day in bed.

'Oh, oh!' said an alarmed voice. 'Don't you think you should get the hotel doctor to see you?'

'Good heavens, no! What on earth for? As long as I stay in the same room temperature and keep warm today, I'll be fine tomorrow.'

I went back to sleep and was awakened, it seemed almost immediately, by the telephone ringing. I looked at the clock which said only half past eleven.

I picked up the telephone. It was Paul.

'I'm just going out for lunch now,' he said, 'and just to let you know that I'll be catching the train to Cherbourg tonight so that I can get the boat back to England tomorrow.'

Used though I was to Paul's unpredictability and apparently irrational decisions, I was incredulous.

'Catching the train tonight?' I croaked out hoarsely. 'But you were supposed to be staying in Paris for another four or five days. And anyway, it would be stupid for me to get up now and start a journey tonight.'

'Yes, of course, Robina dear, but I didn't mean you to come. You just stay there until you're better and then you come back. I've fixed your bill and arranged for you to have any extra money you need.'

'But what about your business appointments? And why *are* you going so suddenly?'

'I'll cancel my appointments when I get to Cherbourg. I'm going because I don't see any reason to stay longer in Paris. I don't want to go out on my own and you won't be clear of fever for three or four days at least and I don't want to run the risk of catching it myself. Well, good-bye, dear, I'll see you at Sutton Place in a few days' time.'

And with that he rang off.

In fact, he did not cancel his appointments so I spent the next two or three days doing my best to placate various irritated friends and business acquaintances who had turned up for meetings with a non-existent Paul, and trying to direct them as to where they could contact him. This was not so easy as Paul's journey did not go as he had foreseen and the last laugh was mine.

The train to Cherbourg did not go there on Sundays – it only went as far as Caen, where he had to spend the night. Although he was pleased with his hotel room, he was obliged to walk there carrying his own cases as there were no porters and no taxis at the station. He had to get up very early in the morning to catch the train to Cherbourg in time to board the *Queen Elizabeth*. However, when he got there, he found the sea was too rough for her to come in to port and that passengers were being taken out by tender, which Paul would not contemplate doing, so he returned to Caen.

Restless, he decided to lunch at Deauville, and then move to Le Havre to wait for the next big ship. There was none on Tuesday, but the *Rotterdam* was due in on Wednesday. However, when Wednesday dawned, Paul having, as usual, consulted newspapers, port officials, weather experts and an available tanker captain, decided that the wind was still too strong. He was then offered the use of a car by a port official to take him the next day to Le Touquet or Calais.

On the Thursday morning, the car duly arrived, the

wind had dropped to a breeze and the sea was flat. As they set off, Paul decided he would like to have a look at the *Liberté* which was in port. He liked it and on the spur of the moment decided to stay on board; he sailed with it at two o'clock in the afternoon. In fact, he had to leave it by tender when he arrived in England and, after being driven back by hired car, eventually arrived at Sutton Place at half past midnight on Friday morning.

He was rather tired when I rang him later on that morning at half past ten.

'Oh hullo, Robina, how's Paris?' he enquired. (Not – how was I!)

'I haven't the slightest idea,' I replied with a smile to myself.

'Why, where are you?' he suddenly sounded slightly alarmed.

'I'm at home in London. I got over my 'flu by Tuesday, so I caught the Golden Arrow train back on Wednesday, had a lovely Channel crossing and was home by half past six in the evening – so I beat you by thirty hours!'

My mother was very angry with Paul over the whole episode and told him so. She pointed out that, as a friend of the family, apart from being my employer, she and my father had trusted and relied on Paul to act 'in loco parentis' when I was travelling with him and she considered he had broken that trust in leaving me when I was ill.

In fact, we all learnt from that incident and others later that Paul, when the crunch came, was incapable of accepting full responsibility when any crisis involving people occurred. Impersonal business crises he could handle with complete assurance but, with human beings, he ran away from making decisions or actually taking the necessary practical steps himself.

However, to give him his due, he could when the contrasting side of his nature was uppermost, be resourceful, understanding and quick to act.

On one business trip to Germany we had an enforced overnight stay at Kaiserslautern, due to having been greatly slowed down by snowy weather and road-repairs. Our visit was an almost unmitigated disaster from beginning to end.

The only hotel we could find with rooms to spare was scarcely up to the standard of the worst commercial traveller's hotel. The bedrooms were long and rectangular with bare wooden floorboards and one worn, dirty, grey runner of carpet down the centre of the room from the door to the windows. I dread to think what germs lingered in the bathroom but luckily we had hand-basins in our bedrooms which I scrubbed out with the disinfectant that I always carried on our travels. I will say the sheets and blankets seemed very clean; Paul looked for fleas, but couldn't find any.

We had to get our own luggage out of the car as there was no porter around. Paul removed the last case and, with a sigh of relief, gaily slammed the door of the Cadillac shut. The only trouble was, he had locked all the car doors and he had also left both sets of door and ignition keys inside the car.

The windows were all shut and were too well-fitting to be easily prised open. Eventually, after much enquiry, Paul managed to get someone to bend a thin strip of metal into an 'L' shape and then, with great patience, he gently inserted this into a very small gap between the side windows and, after numerous failed attempts, which would have made anyone else give up in despair, managed at last to knock upwards and thus release the catch on the window. This meant there was just enough of an opening for me to get in my arm and release the door-lock. Were we relieved! We were also exhausted as this had taken nearly an hour to do, on top of the tiring day's driving.

There was not much choice on the menu, but we were both dying of thirst. We had to make do with what we could get, and so it was the one and only occasion on which Paul and I ever saw each other drink beer.

My bed was very uncomfortable but I was so tired that I eventually fell asleep. Then, towards morning, I woke up suddenly to hear a sound at the door. I reached out a hand and switched on the light, at the same time picking up my fairly heavy handbag to throw if necessary. I saw the door-handle moving slowly and silently upwards – luckily I had, as always, locked it, but I was not too sure that someone else might not have a key that fitted. Suddenly a floor-board creaked outside and the door-handle came down swiftly but noiselessly and all was quiet again. It was ten minutes past six. There was no telephone in the room so I couldn't call Paul; I was too frightened to risk opening the door yet and run down the passage to his room, so I sat up in bed with the light still on and watched the door.

At last, I must have dozed off again, only to wake with a start at another noise; I opened my eyes to see the door-handle moving again. This time it was a quarter past seven so I dropped a shoe (with what seemed a deafening thump) on the bare floor. The handle returned to its proper position. I waited a few minutes but nothing more happened, so I then got up and dressed and immediately felt less vulnerable.

I was all packed and ready to leave that place by a quarter past eight and hoped that, for once, Paul would have woken up. I could hear general noises in the hotel so I gingerly opened the door and peered out. I could see no one so I came out, shut the door, locked it and flew down the passage and banged on Paul's door. Thank goodness, he was not only awake, but up and dressed. For once, he was perceptive enough to see that something was wrong.

'Whatever has happened to you, Robina dear?' he asked, scrutinizing my face.

I told him what had happened and he immediately said he would report it.

'No, Paul,' I said, 'I'm not too sure it wasn't the management themselves – you remember we both thought they seemed a little odd last night – and I've just remembered something else: this is the first hotel where we have not been asked for our passports or made to sign the register for the police. At the moment, there is no proof that we have ever been here and all I want to do is get away as soon as possible.'

Paul agreed at once and, although now apprehensive, insisted we should have at least some coffee before we left to keep us going on the long journey ahead. Even with that, we were out of the hotel and, with relief, away in twenty minutes.

The episode remained so clearly in Paul's memory that he took great pains on all his future trips to work out routes that would on no account run the risk of his ever having to stay again in Kaiserslautern!

Paul had an intense interest in all details connected with travelling, particularly when it was by ship. My mother and I, being avid sea-travellers, were always closely questioned by him on weather, wind force and wave height.

When he travelled by sea, he would spend days beforehand checking the weather reports and forecasts through the appropriate meteorological offices, harbour masters and, sometimes, even the captains of the Company's tankers if they were somewhere near the right area.

When crossing the Channel, he preferred to take the longer Southampton to Cherbourg/Le Havre route on a large ship such as the (old) *Queen Elizabeth* or the *America*. He thought my mother and I, who have always been good sailors, were very brave to risk the often choppy Dover/Calais and Folkestone/Boulogne crossings on (to him) small steamers like the famous *Canterbury* and *Maid of Orléans*. I do not know, and doubt if he did, whether he *would* have been sea-sick on a rough crossing; he was so nervous at the thought of being ill that he never risked being in the situation where it might occur.

When I joined my mother on the annual spring holidays she used to have in New York in the early 1960s, we always went by sea on either the *Queen Mary* or *Queen Elizabeth* as, to us, the crossing itself on those incomparable ships was part of the holiday.

Paul would telephone us in the evening on the first two days after we left Southampton and enquire about the weather, what we had eaten, what films were being shown in the cinema and who else was on the passenger list – in that order!

On our way home, he would again get in touch when we were two days sailing time out and, having first checked on local weather conditions for Southampton (not that we were honestly at all worried), would pass on all the particulars of visibility, wind, tides and any other information he thought useful.

When we went to New York City, we always stayed at the Pierre Hotel which was in those days part of the Getty Group. Although it was supposed to be my holiday, he never lost the opportunity to give me letters and messages to deliver 'as I was there anyway' and, occasionally, even meetings to attend on his behalf. As he pointed out, it saved him the cost of the postage and, possibly, several transatlantic telephone calls.

One of the more droll little errands he gave me one year was to find out why there was such a high outlay in replacements of teaspoons at the Pierre and to cable him with any information I could glean.

After due enquiry, I wired him: 'Customers keeping spoons as souvenirs'.

About five hours later, I received a return cable which said, 'Instruct [the Manager] charge all customers for two spoons per meal'.

When I showed that to the poor manager, he nearly had a fit; I suggested he did nothing in the meantime but waited until I had got back to England and spoken to Paul, as charging the customers for spoons was tantamount to an accusation that they had taken them. In fact, the whole idea was shelved and then forgotten.

When I saw Paul after we had returned, I told him I had brought him one or two mementos from the States and handed him various packages including one box over which I had taken a great deal of trouble. I watched him as he opened it and looked at – three dozen teaspoons from the Pierre Hotel.

Speechless for once, he looked at me in puzzlement, then back at the spoons.

'Read the card with them,' I said.

On it I had explained: 'Meals eaten by two people @ 9 meals each=18

Number of spoons retained @ 2 per meal=36

Herewith, returned, three dozen spoons.'

It was the sort of little joke that appealed to him and he started to laugh – and went on laughing until the tears ran down his cheeks.

He kept the spoons for his private kitchenette at Sutton Place.

Although he was so often such a bad time-keeper, Paul could make the most amazing efforts on occasions.

One year, my mother and I were travelling by the famous Golden Arrow and Blue Trains from London to Monte Carlo, which necessitated spending about forty minutes sitting in a station in Paris while they added carriages and changed the engine and so on.

Paul was then staying at the George V in Paris so he said he would come along to the station and sit and chat in the train with us until it left. I could not believe he would ever get to the station on time, much less the right platform – but he did, exactly five minutes after we had pulled in. It was typical of his kindliness – for the Paris traffic had been bad and in all he spent nearly two hours travelling to come and spend half-an-hour with us, sitting in a very dull station with no refreshments.

It was when we were in Monte Carlo on that trip, that Paul telephoned us one morning. 'I've just been speaking to Ari,' he said, 'and he tells me his yacht has just berthed in the harbour at Monte Carlo.'

'Ari?' I asked, not sure whether I had heard 'Ari' or 'Harry'.

118

'Yes, Ari Onassis. Would you like to see over the yacht? I'm sure I can arrange it with him.'

'We'd love to.'

My mother had often in the past admired the *Christina* and we were thrilled at the chance to see over her.

Half an hour later, Paul rang back to say that Ari would be delighted if we would like to see the yacht and gave us the name of the person to contact.

My mother and I had a fabulous afternoon on the *Christina*. The state-rooms were magnificent – particularly the main salon which had a full-size piano and an open fireplace for burning logs – and the decoration and furnishings superbly elegant.

There was so much to admire and we could not have been more warmly welcomed by the staff and crew who were on board, even to being entertained with a bottle of champagne.

That evening, Paul telephoned to enquire how we had enjoyed ourselves. Not having been on the *Christina* himself, he listened keenly to every detail and, when we thanked him for taking all the trouble to arrange our visit, answered, 'You're very welcome – and thank you for telling me so much about it. I really feel as though I had been there myself.'

When he rang off, I looked at the clock. So much for his frugality. We had been on the telephone for over an hour.

7

Food - but not too much for thought

It was always fun to give presents to Paul: his tastes
were basically simple and he would accept a present
he really liked with a delighted covetousness.

He was particularly partial to fudge, the soft, sugary
sort; he would accept a gift of that with such alacrity
that he almost snatched it out of one's hands. He would
then make a bee-line for the top, right-hand drawer of
his desk in the study, unlock it, slip in the box and
firmly lock the drawer again and put the key into his
pocket. Sometimes one was offered a piece but usually
he would eat it on his own – and it disappeared
amazingly quickly.

He had a very sweet tooth and loved maple walnut
ice-cream, waffles with maple syrup, and endless bars of
chocolate.

Paul was a menace with his chocolate at the cinema;
he always liked to buy a large bar, usually Cadbury's
Milk Chocolate with Nuts. In those days, it had very
noisy grease-proof paper covering the bar, underneath
the outside wrapper. He always seemed to wait until
there was a moment of suspense and total silence on the
screen before, oblivious of everyone else around, slowly
starting to tear off the paper.

We ultimately came to an arrangement. We agreed

that, after he had bought his chocolate, I would unwrap it before we went into the auditorium and put it into a polythene bag (which I had to remember to take with me each time). I would take charge of this, otherwise the chocolate would have rapidly melted when Paul put it in his jacket pocket. A nudge on my arm during the performance meant I had to dole him out a quarter of his bar.

This even happened at film premières and I often wondered whether anyone ever guessed that my pretty but small evening bags frequently held a large bar of chocolate for 'himself', some business letters and the absolute minimum of my own possessions.

Paul had a passion for maple syrup and maple sugar. The syrup one could occasionally find in London but the sugar seemed impossible to get.

When my father used to go to the States on business, he always asked Paul if there was anything he wanted brought back.

'A bottle or two of maple syrup would be very welcome and some maple sugar if you can find it,' would come the reply, and sometimes as an afterthought, 'And a bottle of dark rum, if you can manage it, Tom.'

Whenever friends brought him such presents, Paul would always accept them most gratefully and immediately lock them away in a convenient cupboard or drawer – or hand them into Bullimore's care – to be savoured at some future time, preferably when he was on his own and did not have to share them. It was not really that he was lacking in generosity in not sharing his favourite delicacies, but simply that he enjoyed them so much that he did not want any distraction, like conversation, when he indulged himself.

121

When he was at Sutton Place, Paul never had breakfast at the normal time in the morning; instead, he combined his breakfast and lunch and uninitiated luncheon guests were often taken aback on going into the dining-room to see, sitting on the carefully laid, polished table, amid the flower-filled silver bowls, a packet of Froment wheat germ extract. This Paul would sprinkle over *his* first course of cereal which he would eat, very swiftly, while his guests were enjoying their soup or hors d'oeuvre. After the cereal, he would often have half a grapefruit, also consumed at great speed, finishing just in time to join his guests for the meat or fish course. After that, he would continue with the normal luncheon.

He seldom drank coffee and regarded it very much as an addition to the meal which, if his guests wanted to have it, should be finished as quickly as possible. He did not encourage long, post-prandial discussions over coffee and liqueurs and, if he observed that the other members of the party were not hurrying, he would frequently stand up abruptly and excuse himself, saying he had some papers to read in the study and he would see everyone later in the drawing-room.

If he was in a particularly impatient mood, he would press the bell on the table and, when Bullimore came in, tell him he could clear the table 'as soon as my guests have finished their coffee'. This heavy-handed hint usually had the desired effect.

Paul was also difficult over coffee when we travelled; to me it was an integral part of the meal and, if we were in a hurry, I preferred to have one less course to eat and leave time for my coffee. Once, in a German restaurant, Paul gave me a long lecture about how bad it was for me.

122

'If we are going to discuss things that are good or bad for you,' I retorted, 'there are plenty of things which you eat that are not particularly good for you in that they have high acid or cholesterol contents, so let's agree that you eat what you want to – and I'll stick to my coffee.'

On one occasion when I ordered my coffee, I had intended to have it while Paul was eating his inevitable ice-cream. However, as luck would have it, the waiter did not bring it until after Paul had finished. As soon as the coffee cup was put on the table, Paul made a real scowl and asked for the bill to be brought immediately. It was put in front of him at the same time as my coffee was poured out. He at once paid the bill, looked at his watch and said it was time to go.

I knew perfectly well that he had no immediate appointment to keep, nor even any urgent telephone calls, so I just replied calmly, 'I'm not ready to leave yet. I've only just got my coffee and it's very hot but, if you want to rush off, you go. I'll see you in time for the four o'clock meeting.'

Alone, I finished my coffee and strolled back to the hotel. I had already prepared all the papers for the business meeting that afternoon and, as I expected, he was reading the newspapers.

I sat down opposite him with the files. He lowered his newspaper and looked at me quizzically over the top of his spectacles. 'You got everything there, dear?' he asked.

'Yes – I've checked it all over.'

There was no more mention for the rest of the day about the coffee incident. At dinner, I firmly ordered coffee as usual – not a word from Paul. From then on, I made a point of having my coffee served at the same time as Paul's ice-cream, and there were no further arguments.

The drinks at Sutton Place were, by and large, unimaginative due to Paul himself. He had a good wine cellar laid down by Bullimore but, as he was an almost incessant rum and Coca Cola drinker and really had very little appreciation for the taste of wines, it tended to be a household of whisky, gin and rum drinks.

My mother has always continued the family custom of having a glass of icy-cold champagne on a Sunday morning before lunch and, when we first went to stay with Paul after he had bought Sutton Place, I suggested that he might like to adopt another of our family habits, especially as he had already acquired several others with enthusiasm.

'That's a very good idea,' he said, with surprising alacrity, and rang for Bullimore to tell him to put the bottle on ice.

It was a lovely summer's day so Bullimore brought the bottle and glasses outside where there were white wrought-iron tables and chairs on the stone-flagged terrace overlooking the south lawns. My parents, Paul and I sat back enjoying the view as we sipped from our glasses. Paul looked as relaxed as I had ever seen him.

'I really enjoyed that,' he enthused as we rose to go in for lunch. 'We must do it again next week-end.'

He did arrange it on most week-ends but only when *we* were down there. Sometimes, he would come over to our Kent home or meet us at one of our local restaurants on a Sunday, and would again enjoy his glass of champagne before lunch.

Food would sometimes, contrary to many people's belief of his disinterest in the subject, tempt Paul when he otherwise could not be bothered to go out.

124

J. Paul Getty in jocular mood
drawn by Robina Lund, 1961

J. Paul Getty, King of the
Castle (Combe)

Poor little Rich Man . . .

One Sunday, I was with my family at our cottage which is about two hours' drive from Sutton Place.

'I wonder,' said my mother, 'if Paul would like to come over for lunch. Why not ring him and ask him if he would like to come and have some home-made sausage pie?'

Now, my mother (in the days before she and I became vegetarians), made a superb, mouth-watering sausage-pie of the sort of which you eat twice as much as you thought you could possibly manage. Paul was very keen on her cooking, so I telephoned him.

'Would you like to come over for lunch today?' I said.

'Well, dear, I'd like to but I have quite a bit of mail to go through,' he answered.

'My mother said to tell you we're having sausage-pie.'

'Real, home-made sausage-pie?'

'Yes.'

'Hot?'

'Yes.'

'Then you tell Catherine from me that I'll be leaving in ten minutes and I'll be with you about a quarter of one.'

And he was.

Although Paul always claimed he did not like fancy food and preferred plain cooking – and some of his friends considered him anything but a gourmet – he seemed to have a favourite drink, dish or restaurant wherever we travelled. In Munich, we always had to stay at the Vier Jahreszeiten Hotel, because that was where the Walterspiel restaurant was.

Days before I first went there, Paul was saying,

'Now, Robina, they make a particular dish there and

E   125

I want you to taste it and work out the recipe so that you can make it for me back in England.'

The day of my first visit came at last and we were seated in the restaurant; the menus were placed in front of us. Paul glanced at his, then grinned with pleasure. I did not bother to look at mine – I knew I was going to have my meal ordered for me.

'Tell me,' I asked, 'just what this speciality is.'

'Hummersalat – lobster salad,' he said, triumphantly showing it to me on the menu.

Well, I had to agree with him: it was absolutely delicious and we had it almost every day, savouring each mouthful and discussing the probable ingredients – and I was able to make a passable imitation of it when we returned home to England.

At most restaurants in Germany, other than the Walterspiel, Paul did not spend too long considering the menu. He simply looked down the listed *plats du jour* until he found one with *preisselbeeren* (whortle-berries), and ordered that, whether it was accompanying beef, liver or anything else. He also virtually abandoned his usual drink of rum and Coca Cola, for an 'Ohio' – a very super champagne cocktail which, according to him, was made only in Germany.

Before he moved into Sutton Place and afterwards when he came up to London, Paul for many years used to stay at the Ritz Hotel, and he used it almost as a London office, quite apart from his overnight stays there.

The first time a certain little ritual happened, I was taken by surprise. We were in the middle of a busy afternoon's work at the hotel when Paul suddenly looked at his watch and said, 'Uh, I've got to go out.'

'How long will you be?'

'Not long,' and he disappeared.

About fifteen minutes later, the door opened and in he came with a small square parcel which he was carrying with great care. He opened the doors of a sideboard and gently placed the package inside.

'I'm going to have a rum and coke,' he said, pressing the service bell. 'What do you want, Robina?'

'Well, I think I'll have a coffee, thank you.'

Ten minutes later, rum and coke and coffee set out and waiter departed, Paul opened a drawer and took out a knife ('Saved it from lunch time,' he explained), then carefully brought the apparently precious parcel to the table, opened it – and beamed.

There, exquisitely packed in its box, was a Fuller's Walnut Gâteau! I had one smallish piece and he ate about half the cake, the rest being carefully locked away for later.

Afternoon tea with walnut cake became quite a regular ritual – although, if anyone else dropped in, the unfinished cake was hastily snatched up by Paul and deposited well out of sight in a cupboard or drawer until the visitor had departed.

We were having lunch in his suite at the Ritz one day before driving down to Sutton Place. Paul had had an enormous fillet steak which he had been unable to finish; it was very lean and very tender.

'It seems such a waste to leave it,' he said. 'I think I'll take it back for Shaun,' (Shaun being his favourite Alsatian).

The problem was how to wrap up a rather greasy piece of meat so that it would not leak during the hour's journey. We could hardly ring room service for some greaseproof paper. However, a little ingenuity and searching provided a paper doily from under the

127

entrée dish which, wrapped around the meat, absorbed the worst of the grease; over that we put a clean white handkerchief of Paul's and then parcelled this up in a couple of pages of the (unread) *New York Times* and, finally, the whole package went into a brown paper bag left over from some shopping Paul had done that morning.

Then we only needed to tie it up neatly. We had no string but Paul managed to find some dental floss in what I called his 'jumble-bag' – a small, brown sponge bag that held all sorts of mess (but never a sponge, flannel or soap) and which I swear was never emptied from one year to the next.

With great difficulty, as the dental floss either kept breaking or knotting itself, we managed to tie up the parcel. We collected his case, the remainder of the day's newspapers, various business files and my brief case and set off in the car for Sutton.

We were about half way down when I suddenly had a thought. 'Have you got Shaun's steak?' I asked Paul.

'No,' he said, 'I thought you picked it up . . .'

I've often wondered what on earth the waiter thought when curiosity made him untie that parcel!

128

8

Health versus Wealth

As soon as Paul developed any symptoms he could not diagnose, he would get out his medical dictionary and, with great industry, pore over pages and pages of often totally irrelevant information. He knew that my mother and I had, from wide experience and a mutual interest in medical matters, a good general knowledge of many symptoms, ailments and treatments and he would spend hours on the telephone mulling over with us what this ache or that sharp pain might be. If a friend of his had a serious illness or a major operation, he would worry about the same thing happening to him. He was particularly terrified of diseases which were epidemic or pandemic or which were in any way disabling.

He was a great pill taker and I used to tease him than most ships' doctors carry and his 'medicine cup- sake of trying them out and not because there was any- thing wrong with him. Friends of his in the pharma- ceutical industry in the United States used to send him some of their new products which he received with the glee of a small boy getting a new train-set. He certainly had far more bottles, boxes and packets of medicaments than most ship's doctors carry and his 'medicine cup- board' made quite an extensive dispensary. His personal assistant and I used to throw out-of-date preparations away when he was not around because he would simply refuse to get rid of anything, whatever the date on it.

He tried for many years to persuade me to have anti-'flu injections but I reminded him that every time he had his, he developed a very bad cold whereas I, as a Vitamin C addict, had not had 'flu for years and rarely had more than one cold a year. Moreover, since becoming vegetarians in 1971, my mother and I had each only had one cold.

Paul was very interested in vegetarianism and, as my mother and I gradually gave up eating all meat and fish and re-balanced our diets, he listened avidly to any information we could give him about protein sources, acid content, vitamin requirements, tissue salts and homeopathic remedies.

He followed our diet closely and would ask all the usual questions such as, 'Does it satisfy your appetites?'

'Yes, more than enough – a little goes a very long way and vegetarian food is very sustaining.'

'Don't you feel tired from lack of meat protein?'

'No, there is more protein in several other foods, and less acid content.'

'Doesn't it get dull?'

'No, there is much more variety than being tied down to the conventional old "meat and two veg." routine.'

As the months went past and lengthened into years, he would often grin wryly and say to us, 'I no longer need to ask if a vegetarian diet suits you; you both look younger while I look older. Your skins stay soft and supple while mine gets drier. You've lost your rheumatism while I've got creakier!'

We would point out to him that we had not become vegetarians for medical reasons but for moral ones

130

and that any health benefits were therefore a 'bonus'.

'If I had known as much when I was younger as I do now, I'd have been a vegetarian years ago,' he said one day, 'but I never asked any questions in those days.'

He watched his weight very carefully and, if he considered he was over-weight, he would start, not dieting, but missing meals, usually dinner, sometimes for several days at a time. He was never completely consistent in what he said he wanted to weigh: when he reached 12½ stone or so, he would get quite panic-stricken, saying it would put a strain on his heart, and he would almost starve himself until he was down to 12 stone. Then he would quite often continue not eating until he reached 11½ stone. At one stage, he was insistent that his proper weight was 11 stone but he looked so thin and drawn that it obviously was too light for him. One or two of us told him that it made him look older and he fairly soon allowed himself to get back to 11½ and then 12 stone.

When he first arrived in England in 1959, he was suffering from a little stiffness in one shoulder, a sore foot and he had also developed a swelling – a result, he thought, of some H3 injections. Every day he worried about these, seeing doctors and having massage. Just as the symptoms and causes all got better, Paul decided to go off for another general check-up at the London Clinic because he was convinced that he still was not well. He was told he had a slightly swollen muscle and fatty tissue resulting from his injections and should have short-wave treatment. He continued this treatment for weeks afterwards saying he was terrified that he would otherwise develop a permanently stiffened shoulder.

Having had a hernia when he was younger, he developed recurrent symptoms in the mid 1960s. He discovered there was an injection treatment given in the United States but which, according to my doctor, was not well considered here. But Paul decided to have the treatment – for evermore, it seemed to me. He kept on saying how marvellous it was but when I asked him why he had to have the treatment for so long if it was *that* marvellous, he simply shrugged his shoulders, smiled thinly and said, 'It's my insurance – I feel that as long as I'm having treatment for something, I should stay basically healthy.'

It was the same with massage. He availed himself of the expertise of masseurs and osteopaths at every opportunity and he was always extremely complimentary about the osteopathic skills of Dr Stephen Ward, from whom he had treatment after meeting him at Cliveden.

Before we left on one of our business trips to Europe some years ago, Paul was vaccinated against smallpox. It took splendidly but, naturally, it was very sore and the whole of his upper arm became extremely sensitive. He was petrified, however, that it might have 'gone wrong' or got otherwise infected and I had to inspect it at least twice a day in order to reassure him. I pointed out to him that it was just as well that, had I been able, I would have chosen a medical career in preference to the law and that, because of my interest, I was able to cope with some of the less pleasant aspects of his indispositions. He sometimes consulted me as though I were a combination of nurse, doctor and dietician.

However, if Paul was nervous of doctors and frightened of illness, he certainly had no fears about the

132

dentist. He was a frequent patient, usually expecting to get an appointment on as little as two hours' notice even when it was not an emergency. He did, however, like to have company when he went and I must have sat fifty or sixty times in various dentists' waiting-rooms all over England and Europe.

As he grew older, Paul became more obsessive over some of his ideas and the hardest ones to counteract or cope with were those concerned with his own health.

For two or three years in his middle sixties, he had periods of deep depression and moroseness because he was utterly convinced he was about to die since his father had died in *his* sixties.

'But why,' I argued so many times, 'do you believe you will take after your father? Why not your mother, in which case you will live to your eighties or nineties?'

In the end, of course, I was right but he would not be convinced, not, that is, until the day of his seventieth birthday when he sighed with relief and said, 'Well, dear, maybe you are right. Now I have survived my sixties, I feel I might well live into my eighties.'

One fear of his, sadly, came to pass. He desperately wanted, for many reasons, to die in California. A fortune-teller had once told him he would die 'on foreign soil'.

'I don't want that to happen,' he said several times. 'I want to go home, however ill I am – I'll even fly to get there in time.'

As he got older, he used to say, wistfully, 'I love England, it's my second home, but I'm feeling more homesick every year and when you get old, you want to go back to the familiar things of your youth. You can only ever be a stranger in an alien land – nothing can ever replace the country of your birth.'

He remained in England because of his work. Then he became at first unwell, later ill. Again he said he wanted to go home but he was persuaded to postpone his departure. He stayed – too long – and, indeed, died 'on foreign soil'.

9

Spirits, Stars and Superstitions

Paul was very superstitious – so much so, in fact, that one never dared to tell him any superstition that he did not already have, otherwise he would instantly acquire it and worry over it for evermore.

'But it simply doesn't make sense,' I would argue with him, 'you have lived sixty years and more without knowing that anyone thought such-and-such a thing was supposed to bring bad luck and so you've quite happily never given it a thought. Now someone tells you about it and you at once assume you are going to be deluged with misfortune.'

'But the difference is in the *knowing*,' he would say in a voice of apprehensive gloom.

One day I decided to try to find out his process of reasoning. 'You mean you believe that if *you* know something is supposed to be unlucky, then it is unlucky – but if other people think it's unlucky but *you* don't know about it, then it is not unlucky?'

'Yes, that's right,' he nodded.

'In that case, the answer seems to be simple,' I pointed out. 'Nobody must ever tell you again about any superstition,' and, in fact, most of his close friends and staff never did.

Among Paul's most strictly observed superstitions

was sitting thirteen at table – that he would absolutely refuse to do. Once, when there was a lunch party at Sutton Place, one of the guests was late which meant we were thirteen so Mrs Roxburgh, who was then his Social Secretary, said that she would leave the table to make twelve. Shortly afterwards the missing guest arrived, so we were once again thirteen.

'Mrs Roxburgh must come back at once,' panicked Paul (now standing up, well away from the table).

Meanwhile, the food was getting cold and the other guests were gazing hungrily at their plates.

'You sit down,' I pushed Paul gently, 'and entertain your guests. I'll wait until Mrs Roxburgh comes back and then we will both sit down.'

'All right,' agreed Paul, 'but don't forget that you must sit down together or else it will be bad luck for all of us.'

So when Mrs Roxburgh returned, she and I had to sit down in precisely orchestrated unison, conducted by Paul.

Another superstition, if one can call it that, which he adopted from us was 'first-footing'. As a family, we sometimes used to spend Christmas and New Year at Sutton Place. My mother, being Scottish, had all her life been used to 'first-footing' in the New Year. It meant that to bring luck for the coming year the first person over the threshold after midnight should be dark-haired and come in with a bottle of champagne and my mother, being the only really dark-haired member of the family, was always the one who went out of the house two minutes before the hour struck and was first in afterwards.

The first New Year we spent at Sutton Place, we asked Paul whether he would like to be 'first-footed'.

After my mother had explained it, he was entranced by the idea and rapidly organised a bottle of champagne for bringing in and some more for drinking after midnight. He watched the minutes ticking away and at precisely two minutes to midnight he said to my mother; 'It's time for you to go outside, Catherine dear, you mustn't be late.'

Once she was outside and the rest of us were waiting for the hour to strike, he was as excited as a small boy, continually exclaiming, 'It can't be long now.'

As soon as the first stroke of twelve had struck, my mother knocked on the heavy front door and in a split second Paul had opened it.

'Happy New Year, Paul,' she said, handing him the champagne. He flung his arms round her in an emotional embrace.

'Thank you, darling, and to you,' he said, with a break in his voice and tears were running down his cheeks.

He never forgot first-footing after that. If we were together on New Year's Eve, it became a ritual for him too; it we were apart, we would speak soon after midnight on the telephone and he would always ask, 'Have you first-footed yet? We have, but it's not the same without you all.'

As well as superstitions, the other subject which was officially forbidden in Paul's hearing was that of ghosts – at least as far as their presence at Sutton Place was concerned. Paul always hid behind the excuse that he did not want any ghost stories circulated in case they alarmed the staff, but this lost its validity when, ocasionally, it was a member of the staff who related the story. Ask Paul if he believed in ghosts and he would strongly deny it; then, with a cautious glance backwards over each shoulder, he would lean forward in the dim candle-light and say in a low voice, 'But did you hear what happened to so-and-so when he went to stay with those friends of his in the country?'

And there would follow a detailed account of some logically inexplicable story, occasionally, I think, further embroidered by Paul, who liked to create as creepy an atmosphere as possible and who would relate it in an increasingly sepulchral voice!

He was particularly fond of relating how he once stayed at Woburn Abbey with Ian and Nicole Bedford and how a door in an adjacent room would not stay shut, even after it had been locked. It was, he asserted, a completely unique occurrence.

'That I am sure it is not,' I told him. 'There must be many old houses and other buildings in this country alone where it happens and there certainly are in France.'

'Well,' he said, 'tell me one place in England, then.'

'Right here at Sutton Place,' I said to him, 'but I'm not telling you which door because you don't really want to know.'

'No, I don't,' he mused slowly, torn between intense curiosity and apprehensiveness, 'but you can tell me how often it happens.'

'No, I can't,' I replied. 'To my knowledge, it only has happened at night and, so far, I have only known it to happen four or five times and been told about another three or four occasions by somebody who used to visit the house many years ago.'

He often used to try to get more details from me, but it was only a game in which he indulged because he knew I would never tell him. If he had thought I would disclose anything more, he would never have raised the subject again.

In fact, the door in question was the connecting one between the Oriel Room and the adjacent bedroom, known as the Oriel Dressing-Room.

When my grandmother was alive, she used to stay in that room while I was in the Oriel Room. The dressing-room had two doors, the connecting one and another which gave on to the corridor. My grandmother, like me, was accustomed to locking herself in at night and I usually used to leave her to turn the key in the connecting door.

One morning when I got up early, the door was about two feet open. Alarmed in case my grandmother was ill and had tried to come in to me, I rushed in, only to find her still fast asleep in bed. I retreated quietly, leaving the door as I had found it.

Later, when she was awake, I said to her, laughing, 'You certainly gave me a fright this morning. When I got up, the door here was open. You don't usually unlock it before six in the morning, do you?'

'I never do,' she said, 'and I didn't open it this morning, either. I thought you had.'

'But the key was on your side,' I pointed out, 'so you probably forgot to lock it last night.'

'I might have forgotten to lock it but I would not forget to shut it,' she remarked.

'Well, perhaps it is a loose catch and it clicked open itself,' I suggested.

I closed the door and then pulled at it; nothing would make it open. I inspected the lock but it seemed perfectly all right.

My grandmother, watching me, remarked, quite unperturbed, in her dry cottish way, 'It's not the first time that I know that I've locked it and then found it wide open the following morning and I expect it will happen again.'

'Well, tonight,' I said, 'I'll lock it from my side and we'll see what happens.'

I did and, having turned the key firmly in the lock, I then tugged and pushed at the door but it was quite impossible to open. I put the key on my mantelpiece.

The next morning, I awoke at seven and looked at the door. It was about eighteen inches open, the key was still on my mantelpiece and the tongue of the lock was still extended. Although it was inexplicable, there was nothing unpleasant in the atmosphere and neither my grandmother nor I felt at all alarmed at any time.

Doreen, Lady Brabourne, always had an inexhaustible fund of fascinating ghost stories about Sutton Place, having been a frequent guest in the days of perhaps its greatest elegance, earlier this century. Whenever she could be persuaded by avid listeners to relate some positively eerie story, Paul would either ostentatiously continue eating, drinking or talking or, if none of those were possible, excuse himself on some mumbled pretext.

He could not always avoid the tales of weird happenings, however. One day, one of the craftsmen who was employed in the Long Gallery – I believe he was polishing the furniture or panelling – suddenly appeared back in the main part of the house, white-faced and obviously shaken, declaring that he had seen something near the priest's hole and he was not going back there on his own. Until the work was finished, we arranged for two men to be in the Gallery but nothing further was seen.

My own oddest experience I did not relate to Paul

but the person who shared it with me told someone who did tell Paul. For once, he was sufficiently interested in the historical aspect to overcome his fear of the possibly supernatural. The whole episode, we all agreed, was certainly very strange.

It was in the mid-1960s. I sometimes stayed the night at Sutton Place if I had been working late or had an early start to make and, on this particular night, we had not finished with the telephone calls and paper-work until nearly midnight and, after a nine o'clock start that morning, I was exhausted when I at last staggered upstairs.

In those days, I usually slept in the room known as either the Red Room (although the furnishings were not red but a deep pink), or the Oriel Room, because of the shape of the principal window which, like the adjacent flat window, looked out on to the central courtyard. The room, which apart from paintwork, had not been refurnished or refurbished when Paul moved in, had a four-poster bed covered with a beautiful heavy damask-type cover of material matching the long curtains and the chaise-longue at the foot of the bed. One entered by a door on the left hand side of the room, immediately facing the oriel window; the bed was against the right hand wall, extending outwards towards the centre of the room. There was no con-necting bathroom but the nearest one was immediately across the passage and was, in fact, the original bath-room put in to Sutton Place at the turn of the century (although completely modernised and redecorated since then!).

I always locked the doors to whichever room I stayed in and, although the door to the Oriel Room was in the direct view of the security man who sat all night with a guard dog immediately outside Paul's door, I still felt safer in that large, rather creaking house with the keys turned in the locks.

As it happened, the security man on duty this night was Mr Belton and his dog was my favourite Alsatian out of all those which I have ever known at Sutton Place. He was a beautiful golden creature called Odin, with deep, brown, luminous eyes and with a calm, regal mien; a very intelligent and thoughtful dog.

So, I locked the doors, drew the curtains back a little to let in the brilliant moonlight, climbed into bed and switched out the light. I must have fallen asleep very quickly as I never heard Paul come upstairs although he had to walk over loudly creaking floor-boards in the passage right outside my room.

However, at about a quarter past two, something did stir me into being half-awake and I sleepily opened my eyes. Puzzled, I saw that the counterpane on the bed was not its usual deep pink, but a many-coloured intricate patchwork. Then I realised that the bed was alongside the door, facing the oriel window, through which the moonlight was still shining. I heard a small click in the silence and was instantly very wide awake. I suddenly became aware the door beside me was opening very slowly towards the bed but I could not see who was behind it. I tried to continue breathing regularly and closed my eyes until I was just looking through my eyelashes.

A hand, then an arm, came round the door; caught in the moonlight I could see the arm was covered in what appeared to be an Elizabethan slashed sleeve in emerald green silk and velvet; the hand, which was a man's left hand, had an exquisite, very large, intricately-worked emerald and gold ring on the third finger. The shoulders and face of the man were in the shadows but I could make out enough to see that he was dark and wearing a small, pointed beard.

The figure stood there, without moving. I tried to continue breathing regularly. At last the arm moved slowly forward and the hand reached out to a goblet of liquid on a small table beside the bed. There was an almost imperceptible sound and I saw the emerald top of the ring had opened; the hand turned quietly, palm upwards, over the glass and a fine white powder fell into the liquid.

Then, quickly but silently, the arm was withdrawn out of sight into the dark shadows behind the door which, in its turn, was swiftly but noiselessly closed.

I lay with my eyes shut for what seemed a long time but was probably only a minute or two. I could hear absolutely nothing. Then I sat up and gaped.

The moonlight was still coming in – from the windows on the right of my bed; and there was the door, as it had originally been, over in the farthest left-hand corner of the room. I looked down – at the deep pink damask bed-cover; to my left, there was no goblet nor any small table.

Intensely puzzled, I switched on the light and decided I must have had one of those vivid dreams which, because they seem to have neither beginning nor awakening at the end, become so easily confused with real life and are frequently so hard to forget. Curiously enough, I did not feel frightened in any way, but I decided to get up and walk about and also to have a splash of cold water on my face to make sure I was properly awake.

So I put on a fairly voluminous housecoat, unlocked the door and stepped out quietly into the passage. As

I crossed over to the bathroom, I looked down the passage towards Mr Belton and Odin and waved. Odin, sitting alert and upright, watched me with ears pricked; Mr Belton sat looking but did not move.

Having refreshed myself with plenty of icy-cold water, I felt much better and, as I crossed the passage again, I once more waved at Mr Belton and Odin. Odin wagged his tail gently, suddenly seemed to relax and lay down. Mr Belton leant forward slightly but did not return the wave. I wondered if, as I was at the more dimly-lit end of the passage, he had not in fact seen me and I made a mental note to check it for myself the following evening.

I went back to bed but, I have to admit, left both the light and the radio on. I fell into a sound sleep and didn't wake again until the maid brought early morning tea.

Because of his fear of anything remotely akin to the supernatural, I said nothing to Paul. Mr Belton, of course, was off duty by the time I appeared in the morning. However, to my great surprise, he came to the office in the afternoon and asked if he could have a word with me in private.

'Yes, of course,' I said, wondering what this could possibly be about. When we were on our own, he took a deep breath. 'I wanted to apologise to you, Miss Lund,' he said. I looked at him with surprise.

'You see,' he went on, 'I didn't mean to be rude to you last night; you remember when you came out of your room and crossed the passage to the bathroom and then came back and each time you waved to me – you did, didn't you?'

'Yes,' I said, 'and you didn't wave back – but I thought you probably hadn't seen me – Odin did.'

'Oh, I saw you all right.' He nodded pensively. 'But it was because of what I saw before that I didn't.'

'What was that?' I asked, alert.

'Well,' he said slowly, 'it was most odd, really. A few minutes before you came out, Odin growled softly as he looked along the passage and he and I peered down. I thought I saw a figure, a man dressed in old-fashioned clothes, stop outside your door and then go in; after a moment or two, he came out and then just disappeared. Two minutes or so later, you came out. The housecoat you were wearing with the big sleeves looked as though it could have been one of those dresses ladies wore in old oil paintings, so I thought I was having a dream and you were still part of it. The second time you waved, Odin relaxed so I knew it was you, but by then you were back in your room and it was too late for me to wave back. Anyway, that was what happened to me and I'm very sorry if I seemed to be rude.'

'Don't worry about that,' I said, 'but I would like to ask you more details about your "dream", if that is what it was. What sort of old-fashioned clothes was the man wearing?'

'Oh, I know that from the paintings,' he said confidently. 'Elizabethan. He wore doublet and hose – you know those baggy trousers like plus-fours which come to just below the knee – and a jacket with big sleeves.'

'What colour?' I asked.

'A beautiful brilliant green,' he said.

'Did you see what he was like?'

'He was about 5 feet 10 inches, dark-haired and with a little dark, pointed beard and a moustache curling upwards.'

'Well!' I shook my head. 'I don't understand it but I will tell you what happened to *me*,' and I related my 'dream'.

'Have we seen a ghost?' I wondered. 'It's all very strange. Have you ever seen this man before, Mr Belton?'

'Yes,' he admitted, 'once or twice, but it was very funny last night. You see, he usually goes into the Oriel Dressing-Room, the next room to yours. I've never seen him go into yours before.'

We agreed we would say nothing either to Paul, because of his nervousness, or to any of the staff, some of whom might, understandably, have been frightened. Mr Belton had already told a friend his part of the story but hoped he would keep discreetly silent. I told only my parents. In fact it was not until two or three years later that Paul came to hear of the story at all – from an estate tenant who presumably had heard it directly or indirectly from Mr Belton's friend. By this time, however, several other people (including my father) had slept completely peacefully in that room, so Paul was able to take a more detached view of the story.

He still did not want it spread about, which I quite understood, but he was sufficiently intrigued to read all the old books and records about Sutton Place he had collected without, however, finding any trace of a story which would account for such a ghost.

Some time afterwards, we went to a dinner party in London at which Doreen, Lady Brabourne, was another guest. Paul, knowing only too well her interest, related the story to her – with an uneasy laugh about the whole episode.

'You know, Mr Getty,' she said, 'that in Lord Northcliffe's time that room was supposed to be the most haunted room in England.'

Paul's mouth dropped open and he swiftly emptied his glass of brandy.

'If you worry so much about ghosts, do you believe in reincarnation?' I asked Paul one evening soon after he had learnt about the Oriel Room visitation.

'Yes,' he replied slowly, 'I think I probably do. I know it isn't recognised by the Christians and Jews but I think the Eastern religions have a great deal to offer to people who want to think deeply about their beliefs and not just accept, without question, what someone tells them from a pulpit. I was brought up a Methodist but I've been to services in churches of many denominations and I can't honestly say I see many differences between them except in matters which I consider trivial. I don't approve of churches and clergy which are over-resplendent with gold ornaments and wonderfully embroidered robes in parishes which are poverty-stricken. I don't believe any religion has the right to claim to be the one true religion; I don't believe in teaching children to fear God. I don't accept the right of any Church to teach that the followers of any other Church are wrong. In fact,' he said with a thin, wry smile, 'while I believe in the theory of religion, I guess I don't believe in its practical application. How much destruction and how many terrible deaths have been caused over thousands of years with religious beliefs as the excuse, when the truth has almost always been a lust for power and possessions.'

He paused. 'We seem to have wandered off the original subject,' he remarked, looking puzzled.

'We usually do,' I said. 'One thing always leads to

another in our conversations. But, returning to rein-carnation, why do you believe in it?'

'I suppose for two reasons. The first is that it ex-plains why some people seem to know facts or places from the past; it's as though their memories have been transferred to a new being. Secondly, it's a sort of in-stinct I have. I have always felt I had a great deal in common with two people widely separated in time – Randolph Hearst and Hadrian. I can scarcely be a reincarnation of Mr Hearst with his being a contem-porary of mine, but I have wondered for many years why I have for so long felt such a close affinity with Hadrian. When I read about him and his villa and his life, I feel I already know it all and understand why he made the decisions he did. I would very much like to think,' he added wistfully, 'that I *was* a reincarna-tion of his spirit and I would like to emulate him as closely as I can.'

It was some years later that he finally agreed the plans and put in hand the building of the J. Paul Getty Museum in California – based on the design of Hadrian's Villa at Herculaneum.

'What about Randolph Hearst?' I asked. 'In many ways he seemed quite the opposite to you. He liked to spend his money in a grandiose fashion, which one can't really accuse you of doing – except when it comes to art!'

Paul laughed. 'No, I guess I'm a bit more close-fisted than he was but I ended up richer! The funny thing was, I wanted to buy San Simeon but I thought it was too isolated for convenience, so I ended by buying Mr Hearst's two fine long dining-tables from his castle, St Donat's, and I think they look better in my dining-room here at Sutton Place. But, you know,

Hadrian, Hearst and I are alike – we have all liked things on a grand scale. Palatial buildings, fine pictures, gold and silver plate. We even all share a love of swimming pools although my two at Sutton Place are very modest compared with those that they had. We've all been collectors of objets d'art of many types, even if our tastes have not always been quite the same.'

'Is there anyone else you feel you might have been in an earlier life?' I enquired as he began to go off into a contemplative muse.

'If I had any ability to ride a horse, I'd say Alexander the Great. I'd like to have been Leonardo da Vinci – he was probably the greatest designer and inventor of all time – but I'm sure I wasn't! I sometimes wonder if I lived in the Italy of the Borgias – I admire their art and furniture and architecture so much, to say nothing of their diplomacy and politics. I suppose if I were asked what country I felt I belonged to in the past, the answer always would be Italy.'

Now, in retrospect, it seems a particularly ironic twist of fate that, towards the end of his life, it should have been a series of events in, and associated with, Italy that prevented him from fulfilling his ambition to live there and embittered his previous fascination and, in some respects, awe for the land and its peoples through so many centuries. The bizarre circumstances began simply enough.

'My ideal, leaving aside business commitments,' he had often said, 'would be to arrive in England at the beginning of April, in time to see the thousands of daffodils on the south lawns and the trees beginning to come into leaf. I would leave at the end of August and spend a good part of September in Scotland where I like the scenery, the people and the food then I would go and winter in the warmth of Italy and laze away the months just swimming and sightseeing at the old archaeological sites like Herculaneum. On my way back to England, I would drive a circuitous route calling at the main European museums and art galleries in order to see their latest acquisitions.'

One week-end when my parents and I were staying at Sutton Place, Paul announced that he had decided to buy Palo. A few miles south of Rome, it was a palatial villa which he acquired from Prince Odescalchi, together with the lease of the adjoining small castle and keep.

We were about as surprised at his decision as we had been some years earlier when he had telephoned to say he had agreed to buy Sutton Place. He explained that he intended to live at Palo each winter and hoped we would join him there.

'Is it going to be practical?' I asked. 'What are you going to do about staff? Are you keeping a full staff at both Sutton Place and Palo?'

'Oh, no,' he said, 'I shall take the Sutton Place staff with me and when we come back to England I shall just leave a caretaker couple in charge of Palo.'

'Suppose the staff here don't want to be away for as long as that? Have you asked them?'

'No.' He looked surprised. 'I haven't asked them but I can't see anyone objecting to six months' vacation each year in Italy.'

'But, Paul,' I protested, 'you *must* ask them. For one thing you are not inviting them over for a six-months' holiday, but to carry on with their usual work; secondly, they may not want to go at all because of their own personal circumstances and, thirdly, they may only want to go for a month or two, but not for as

150

long as six. They may work for you but they were all engaged to work in this country and you are not paying them to work for you twenty-four hours a day, seven days a week – even if they frequently do! You will be affecting their time off as well as their working hours and you must not expect everyone to be instantly agreeable. What suits you doesn't necessarily suit them.'

'Well,' he conceded, 'I guess not. Anyway,' he brightened, 'that's a detail we can work out later.'

I did not pursue the argument further at that time.

'I've got some fine colour prints,' he said. 'I'll show them to you.'

They were extremely good photographs, very clear and taken from all angles so that he was able to point out where he was going to repair this and repaint that and which windows belonged to which rooms.

Having by now lost a lot of interest in Sutton Place, in which the major part of the repairs and restoration had been completed, he turned his full enthusiasm on to restoring Palo. However, he had the drawback of supervising the greater part of the work at long distance from Sutton Place. He used his various friends who had considerable knowledge of Italian antique furniture, to help and advise him in furnishing the place and to do the interior decorating. Any other friends who were either going to, or returning from, anywhere in the vicinity of Rome, were commandeered ruthlessly by Paul to take messages for him or bring back photographs, patterns, or even just comments and descriptions of the progress of the work.

I felt instinctively that Paul was about to make another big mistake but I tried to respond to his enthusiasm. I noticed my mother was certainly not enraptured by what she saw and, as she never pays false compliments or says what she does not mean, was covering up her inability to offer him encouragement in his new project by asking a lot of questions. He was delighted at her apparent interest and was, as usual, not perceptive enough to notice our lack of whole-hearted contratulations.

Knowing my mother's infallible Scottish gift of second sight, I was sure that she had sensed something and it did not augur well. It was all I could do to wait until we were alone, well out of earshot, in her bedroom, to ask, 'Well, what do you think about it?'

'I don't like it at all,' she said decisively. 'It has a bad atmosphere and it will bring him nothing but misfortune. And I will tell you something else,' she added, 'he will never live there. He may stay a week or two on two or three occasions but that is all.'

She was completely right. Through a strange combination of events – strikes, weather and business commitments – Paul never stayed in the villa for longer than about three weeks at a time and then only on three or four occasions. Finally, when his grandson Paul III was kidnapped, it was the last straw and Paul declared vehemently that he would never again return to Italy and that it had brought little but misfortune to himself, his son Paul II and to Paul III.

It has always seemed such a waste that my mother could have warned him against Palo, and might thereby have saved him time, money and a bitter disappointment, but we were unable to do so because, once more, of his fear of anything he regarded as having to do with extra-sensory perception.

It was not very long after Palo that he bought a small private island with its own villa just off the coast of Naples. Once more, when my mother saw the photographs, she pronounced that Paul would never even have a holiday there.

As far as I know he never slept a single night there.

My own feeling, when I had seen the photographs of Palo, was of a sort of 'creeping of the skin', especially when I looked at one part of it. I don't know to this day if anything had ever happened there in the past but I did know at once that nothing would ever induce me to visit the place, much less to stay in it; so, when, later, Paul asked me to be a trustee of the property and said that I and my family could stay there whenever we liked, I refused, as politely and gently as I could. Since I was no more able to tell him the whole truth than my mother, I merely explained that, for my own personal reasons, it was extremely unlikely that I would ever stay in the villa; consequently, I felt it was inappropriate to have an 'absent' trustee and that it would be more practical for him to have someone who could keep a regular supervision over the estate. He wasn't very pleased at failing to get his own way but accepted it was common sense.

A curious postscript to the purchase of Palo occurred some months later. Paul and I went to a film première of one of those biblical epics that were for a time so fashionable. The opening scenes were of the Creation which were quite cleverly done, some effects being very impressive. Then the scene changed to represent the Garden of Eden. Sunlight shone through dreamy green trees and Adam awoke from a bed of long grass to the accompaniment of suitably dulcet music. The camera wandered through what seemed a never-ending series of dells and woody groves and I began to feel fidgety and ill-at-ease.

153

Soon I found the atmosphere around me seemed to be getting increasingly hot and oppressive. I looked round but everyone else, including Paul, had eyes to the screen. When I had looked away from it, I felt better; when I looked back, I instantly had a sensation of great dread. It was so extraordinary that I tried to concentrate on why the film was having such an effect. I came to the conclusion that the scenery seemed to be primeval – and evil. I was struck by the absence of any natural sounds or movements; no birds singing, no occasional butterflies, no wind in the trees, not even a rustle of a branch.

The scene moved on to a sort of clearing in the centre of which was a large rounded tree, what type I did not know, hung all over with white flowers. I can only say that, to me, it was like a symbol of all the things that could, or had, ever frightened me, collected into one object. I was perspiring and I knew that if I didn't get out into the fresh air quickly, I would pass out in a faint – an unheard of thing for me to do. Trying to contain the beginnings of a panic – making my way out from the centre of the row so soon after the film had begun and with so many important guests there, was going to be a nightmare in itself – I began to lean towards Paul, hoping I could explain rapidly in a whisper, that he would hear me and maybe even come out with me.

Just then he turned towards me and, with pride in his face and voice, whispered, 'How do you like the scenery? Wonderful, isn't it. It's all shot in my garden at Palo.'

He did not wait for an answer, which was just as well as I was sitting looking at him with my mouth open.

By a really fortuitous bit of timing, the scene changed completely and my relief was so great that I instantly felt slightly better. Ten minutes later, with the Garden of Eden left far behind, I felt perfectly all right again and never had another qualm during the remainder of the film, apart from an effect of slight shock.

In retrospect, I came to the conclusion that it was something malevolent that seemed to emanate from the screen. In infinitely milder form, I had had the same reaction when looking at those photographs of some of the buildings at Palo. Before the film, I had not even known that the garden existed, so I certainly did not know that any filming had been done there, nor could I ever have guessed that the garden on the screen was in any way connected with Italy, let alone Palo. I certainly have no fear of trees – in fact, I love to be among as many as possible, and particularly very old trees.

It was a subject which I could never broach with Paul and I have never had the chance to solve that enigma.

One day, early in 1964, when I was browsing at a bookstall, my eye was caught by a magazine on astrology, advertising on the front cover an article about Paul. Intrigued, and rather sceptical, I glanced through it. Further intrigued and impressed, instead of sceptical, with what I read, I bought the magazine.

My contacts with astrology being limited to reading 'What the stars foretell' in one newspaper or another from time to time, the name of the author of the article, Peter J. Clark, meant nothing to me and I was absolutely certain that he and Paul had never met nor had any personal contact, Paul, as I have already said,

155

being nervous of sixth senses, second sight and other gifts which he could not comprehend.

I mention this because the article was so astonishingly accurate in its assessment of Paul. Even now, when I think back over all the articles, interviews, biographies, and 'autobiographical' précis produced about him by people who had the opportunity of meeting him, speaking to him and appraising him, there has not been one even half as accurate as this astrological assessment. Since there was nothing in it which would in any way have alarmed Paul and I was sure he would be interested in it, I showed it to him.

He perused it slowly and carefully from beginning to end – three times. Finally, he looked over the top of his gold-rimmed spectacles.

'Remarkable!' he said. 'I'm tempted to ask Mr Clark to give me a rundown on all the Company's employees and I think he'd be a great deal more accurate about people than some psychiatrists.'

By one of those strange coincidences, I met Mr Clark at a charity reception a few weeks later. Naturally, I was most curious to speak to him and we retreated to a quiet corner where we would not be overheard. Having complimented him on his article, I could not resist enquiring, 'Is there anything else you did not put in your article?'

'Oh yes,' he replied, 'quite a lot – but what with lack of space and the fact that we must always be careful not to "foretell the future" or to upset people, one is limited in the amount that can be written. However, what I will tell you, in confidence as his lawyer, and in case you may be able to protect him, is that he has a very serious time ahead of him during the next decade or so. I did not stress this in the article and you must not tell him because I know he is highly superstitious and almost neurotically frightened of bad health, bad news and death and all three are going to affect his life greatly.'

I promised that I would say nothing – with my own knowledge of Paul, I would not have done so anyway – and I kept my promise. I never even mentioned that I had met Mr Clark in case it led to Paul asking any further questions which might have been difficult to answer.

However, I certainly remembered those prophetic words as I watched him over the succeeding years, struggling with so many crises, some small and too many big as, for instance, the tragic death of his eldest son, George, and the kidnapping of his grandson.

Personal Tastes - Sweet and Soured

In the early 1960s, Paul considered that his son Paul II should have been a university don and that he was not cut out to be a businessman.

'Paul,' he used to say, 'should live a life of quiet thought and research, surrounded by books. He lacks the single-mindedness and determination to be a businessman. In some ways he reminds me of John D. Rockefeller III – they are both very humane beings.'

Ronald, on the other hand, seemed to veer too far in the other direction for Paul's liking.

'I remember taking Ronny to the Louvre some years ago,' he would relate, 'and while I was admiring several of the exhibits, Ronny casually sauntered through the galleries. I don't reckon he could have told you one exhibit that we passed; I guess he's too much the businessman and no real lover of art.'

Paul also used to complain that none of his sons was willing to work as hard as he did.

'They seem to lack drive,' he said. 'I'll work a twelve- or fourteen-hour day, every day of the week, if there's something that needs to be done; they work very hard for an eight-hour day, then they're off pursuing other interests. I've told them that's not how to make money.'

'Perhaps they're happier, though?' I suggested.

He threw me a look over his glasses – but did not answer.

The friendships I had made over the years with various branches of the press were tested to the full at the time of the kidnapping of Paul's grandson who, for ease of identification, we always referred to as Paul III.

Grandfather Paul was very nervous during the months of the kidnap and did not want any telephone calls connected with it, or any press enquiries, put through to Sutton Place and asked me to handle them. Since he was very upset at the time, particularly with this coming so soon after the death of his eldest son George, I did – but rather unwillingly. Paul, who with a typically self-centred attitude did not want the upset and inconvenience coming into his own home, was quite unperturbed that it should come into mine, disturbing not only my peace but also that of my parents with whom I live.

Although he said the calls were occupying too much of his office staff's time, despite the fact he had four people who could answer the telephones, he expected me to cope with everything on my own. The answer was, of course, that I did, thanks to the endless support and help I received from my mother in particular who, in spite of poor health and failing eyesight, took messages, remembered names and addresses and detailed facts and kept me going with coffee and food. For over six weeks, I often had little more than four hours' sleep during the night and perhaps two or three hours off during the day when I went out for a break and shopped or walked our dog.

When Paul rang up for news, I would tell him about how many telephone calls there had been or that reporters were sitting in their cars outside my home waiting for news and that I could not go outside the door without being surrounded and interrogated. He would just remark 'Oh my!', pause, and start some new topic of conversation.

It was not just that he was not interested or unfeeling; he was so unimaginative by nature that he could not conceive of a situation that he himself had not experienced. It was why he so often reacted to other people's problems with such a lack, as the case might be, of anger, sympathy, concern or even interest.

On the other hand, he could get quite obsessional and, once something attracted his attention, he became totally preoccupied with it. The same single-minded purposefulness that he used in business was brought to focus on his current interest whether it was a painting, space travel, a book, a piece of furniture or anything else.

When his painting, the Madonna de Loreto, was discovered to be by Raphael and not just an unknown Italian-school painting which he had bought at a saleroom years before for under £50, he, understandably, got enormously excited and set his mind on how to find out its background, who had owned it, when and how it had come to be 'lost' and then offered for sale in England quite openly, its true nature unrecognised by any of the art experts. He related constantly the progress of his enquiries and the results of the research that he had various people working on for him. It was very interesting at first but when it went on week after week and month after month, some of his friends remarked that they felt like screaming every time the name Raphael was mentioned!

In the same way, when he was totally immersed in

160

buying some of the gold and silver plate for Sutton Place, my parents and I went to tea with the then Governor of the Tower of London and took Paul with us. We had tea in the private apartments, saw the Crown Jewels, inspected the impressive armoury, were taken on a specially conducted tour of the whole edifice, watched the ravens and absorbed the historical atmosphere as twilight descended silently over the stone walls.

Afterwards, we asked Paul what he had liked best during his visit.

'That solid gold sixteenth-century salt cellar,' he said.

He had some unexpected fears for a man such as he. My parents and I were sitting one evening in the lounge of the Hotel Splendide in Lugano, having our after-dinner coffee, when a quiet, deep, American voice said from behind us, 'May I join you?'

After Paul had settled into the low armchair and we had laughed about the coincidental meeting, he asked, 'Well, and what have you done today?'

'We went up San Salvatore,' we replied (San Salvatore being the local ascent, famous for the extensive views from the top over the northern plains of Italy).

'To the top?' He sounded astonished.

'Yes, right to the top – and we got stuck half-way up!'

'You went by the funicular?'

'Yes.'

'Oh my, oh my!' he exclaimed, his mouth open, eyebrows raised and eyes goggling. He shook his head slowly. 'I think you really are brave – nothing would get me on the funicular.'

And, for years after, whenever the conversation

161

turned to anywhere mountainous or near Switzerland or Italy, he would say, wonderingly, 'Do you remember when you went all the way up San Salvatore in that funicular?'

Although, by and large, Paul did not react emotionally to many things – his feelings usually being limited to 'it's worth having' or 'it's not worth having' – there was one occasion on which he was in as much of a hurry to leave as I was.

We had gone to see a German museum of porcelain ware, most of which I found extremely dull as there seemed to be room after room filled with white or cream china, unrelieved by any colour. Then we came to a few rooms decorated with wine or crimson velvet walls and curtains and very overpowering high four-poster beds. These rooms led on to a strong-room containing some saintly relics; this room, too, was decorated with wine-coloured velvet and had a very hot and airless atmosphere in order to preserve the remains.

Paul and I found the whole display quite revolting – and he was actually out of the door faster than I was!

'I'll never come to this museum again,' he swore. 'People should be warned in advance about exhibits like that in case they don't want to see them.'

Our business tour through Europe in the autumn of 1960 developed into a concentrated cultural tour as well. Apart from castles, museums, art galleries and various other historic buildings, we went to theatres and operas wherever we stayed.

Paul, while not very musically knowledgeable, had a great thirst for the atmosphere and excitement of the occasion. We saw *Othello* and *Volpone* (in German) in Munich and *Der Rosenkavalier* and *Don Giovanni* at the Vienna State Opera.

I was bewitched by the interior of the Opera House, with the high ceilings and five tiers of boxes and galleries but Paul spent all the intervals peering through his opera glasses at the people who were walking up and down, looking and waiting to be looked at, in case he recognised anyone; he was quite downcast that he did not!

As we travelled across Europe, we must have seen almost every major museum, art gallery and schloss on the way. Paul's interests and mine did not always coincide and, at first, we had to go off in precisely the direction *he* wanted but, when we reached the exhibits which interested me, it always seemed to be time for the place to close or for us to get back to the hotel for some business meeting or long-distance telephone call. We eventually agreed that we would each go our own way, meeting at the main entrance again in half-an-hour or an hour's time.

This worked until the day that Paul finished his part quickly and departed, claiming later that he could not find me and presumed I had already left. When I expostulated that *I* would never leave without *him*, he merely said, 'Oh well, I'd seen all I wanted to, anyway.'

'But I hadn't,' I retorted, 'and I wasted half-an-hour looking for you in the gallery.'

He grinned, quite unperturbed – but flattered!

The next day, we went to another museum. There wasn't much there that interested me so I finished my round of viewing early and, as Paul was not waiting by the entrance, I deliberately followed his example of the previous day and left.

I waited for him at the hotel. When he arrived and saw me, he was quite upset and immediately started relating how he had searched all over the museum for me and had got quite worried.

'Now you know how I felt yesterday,' I told him, 'so remember from now on, don't just walk away and leave me in a strange place when we have arranged to meet.'

And he never did again.

Paul was fond of animals up to a point – that is, in much the same way as most people who claim to be so but, in fact, never think very deeply on the matter. His liking for them was basically subjective – for his own enjoyment or pride of possession – but he really understood neither them nor their needs.

In one respect, however, he was determined. He hated hunting and shooting in his later years, particularly after he had settled in at Sutton Place and became used to seeing the beautiful pheasants sauntering through the undergrowth of the newly planted wood or the hares that ran through the long grass and suddenly sat upright in the middle of a bright field of daffodils.

One acquaintance of his, who was of the 'huntin', shootin' and fishin' ' type and who used such parties to acquire clients for his business, had been accustomed to go shooting on the estate in the 5th Duke's time and once asked Paul if he could continue to come down for the odd day's 'sport'. Rather unwillingly, Paul agreed, but later became less concerned about it upon discovering that the gentleman in question was not a good shot and the only thing in danger of being hit was the gamekeeper who accompanied him!

As the park surrounding the house was completely

bisected by the drive, which ran from the London Road gates past the big house to the Woking gates, most of the wildlife crossed the road quite frequently. Paul would always drive the Cadillac slowly, particularly along the stretch from the weir to the house, where the old wood was on the right and his own newly planted wood of fir and oak on the left, and was really excited each time he was rewarded with the sight of the squirrels playing 'last across' or a pheasant looking back down its tail at him with obvious annoyance at its peace being disturbed. He was as wide-eyed as an unspoilt child the first time he saw a weasel streak across the road in front of us and clutched my arm, whispering, 'What's that? I never realised they could move so *fast.*'

A bone of contention between Paul and many of his friends and staff was the question of the lions which he kept at Sutton Place.

When he lived in California, he used to own a lioness, Teresa, whom he allowed to roam around loose on his estate and of whom he always said he was very fond. However, when he left on one of his protracted European tours, he gave her to the local zoo, which may have suited his convenience but which must have been an unhappy move for the animal who had been used to personal care and attention and freedom.

So it was that, when once Paul was given a birthday present of a lion cub (in spite of so much publicity about the unsuitability and cruelty of giving wild animals as presents), there was a concerted effort by many of his friends and acquaintances – including John Aspinall, an expert on the 'big cats' and particularly tigers – to persuade him to give the animal to someone who already owned a pride of lions which

165

roamed free in a park and into which the cub could be introduced. This would have meant it would have been among its own kind – and uncaged.

But, unfortunately, Paul was dazzled by the acquisition of the 'king of beasts' and the symbol of power and strength and invincibility. He was told by the experts that once a wild animal has the 'imprint' of a human on it, that is, has been handled and petted by humans and become used to them, it is virtually impossible ever to return it to its own kind in the wild or even in captivity; the animal will neither settle down itself nor be accepted by the others. He was also told that modern enlightened thinking considered it cruel to keep an animal on its own without a companion of its own species. However, Paul, as so often happened when confronted by tricky decisions which he did not want to take, dithered and delayed until the months had passed and it became too late to transfer the adolescent lion to a strange pride. At last, we were able to persuade him to get a companion lioness and to build them both a much larger enclosure where they could get a more reasonable amount of natural exercise. He was delighted when they had cubs but, even then, it was still from a sense of personal gratification without any balancing acceptance of responsibility for them – for neither in his lifetime nor in his will did he make any provision for them or leave any instructions as to what was to become of them after his death.

It is to be hoped that the new legal requirement that licences must be obtained in order to keep certain wild animals on private property will prevent such stupid and thoughtless presents being given in the future.

The Alsatian guard dogs at Sutton Place achieved

at one time a considerable notoriety. Having been five at first, they began to increase in number fairly rapidly and the offspring brought the total, after a while, up to seventeen. The result was several half-trained, half-controllable, adolescent and excitable dogs who were apt, if out on the loose for exercise without someone who really had some authority over them, to take a bite at the odd visitor or guest who happened to be around. Even the head security man once got badly bitten by Paul's favourite, Shaun, who was a handsome, massive dog – overweight most of the time due to being fed extra tit-bits between meals by some of Paul's women friends who hoped thereby to get in, or keep in, favour with him.

At one stage, it got to the point where, if anyone turned up at the local hospital with a dog bite and asked for an anti-tetanus injection, he or she was automatically asked, 'Sutton Place?'

Latterly, with more security staff trained as dog handlers and with fewer dogs, the animals were kept more under control but those of us, whether friends or staff, who had dogs of our own, remained chary of having them around anywhere near the Alsatians, as both pet and owner were liable to get attacked.

Paul's friend, Penelope Kitson, frequently used to take her Norwich terriers to Sutton Place and they seemed to be accepted by the Alsatians, probably because there was nearly always at least one Norwich terrier running around as Penelope gave Paul one, called Winnie, as a present. While he said he was very fond of Winnie, he never seemed particularly at ease with small dogs since, like some people with children, he just did not know how to treat them. He certainly could have astonishing lapses of incredible stupidity as, for instance, the time he was driving Penelope in the Cadillac on the estate with one of the Norwich terriers 'having a run', as he put it, alongside the moving car. He was apparently incapable of foreseeing what, not surprisingly, happened – that the little dog swerved sideways under the wheels and was killed.

Paul was a tough businessman with whom to undertake any negotiations. Oddly enough, although some of the high-powered company presidents, directors, executives and experts with whom he had meetings were very astute and intelligent, very few realised one significant characteristic of Paul's: he had an immediate sense of rivalry with any other man so his instinctive reaction was to concede nothing or, at any rate, as little as he could; to do more, he felt, would appear as a sign of weakness.

With women, on the other hand – even women in business, provided they were well-qualified, smart, intelligent and had suitable experience – he was far more flexible. He found it almost impossible to separate their femininity from their professional attributes and so reacted as the chivalrous 'gentil-homme' of old, and would listen with infinitely more patience and a far more malleable attitude to their arguments. He liked to agree with them if he possibly could for the simple reason that he felt that, contrary to the male reaction, it raised him in a woman's estimation, which flattered his ego.

If only more men had realised how deeply ingrained these attitudes of Paul's were, they could have saved themselves a great deal of time, trouble and, possibly, face. As it was, a few businessmen and journalists did appreciate the mollifying effect of a woman's presence

168

and used their knowledge to considerable effect. One journalist, in particular, rarely turned up without an attractive 'secretary' in tow; there was usually a different young lady on each visit and she could seldom take any shorthand but it did not matter as long as she was attractive and pleasant. Paul would answer all questions for her benefit and be polite, tolerant and charming.

All his life, he could hardly ever say 'yes' to a man or 'no' to a woman.

'To them that hath shall be given,' was repeatedly proved by Paul's experiences. Although he could well afford to buy almost anything he wanted, he was always being presented with complimentary gifts which most people would love to have but could not afford to buy for themselves.

One example was the electric golf-car which was given to him by an American friend. Run off a battery, it could seat four or five adults, was painted white, with a white canopy over the top and was simplicity itself to drive. Paul did not play golf and, although there had been a nine-hole course in the park earlier this century, it no longer existed. So, after his initial outings in the little car, Paul did not bother with it again – it never seemed to occur to him that he could use it on a fine day for just gently meandering around the estate. Some of us would get it out, however, and cajole him into coming for a ride, then he would thoroughly enjoy himself, driving along yodelling or Indian war-whooping at the top of his voice.

He was occasionally presented with a complimentary overnight stay at a hotel if he had gone there for a well-publicised function, although one such visit did not have the effect intended by the management, namely, to encourage him to stay there more frequently.

We had been to another of Maggi Nolan's splendid parties at a West End hotel and arrangements had been made for Paul to stay there for the remainder of the night instead of returning as usual to the Ritz Hotel.

The following morning, when I picked him up to drive him back to Sutton Place, I asked him, 'How did you enjoy your suite at the hotel?'

'It was really glamorous,' he answered. 'In fact, it was gorgeous and the bed was quite good and comfortable and I slept from the minute I put my head on the pillow until you called me at eleven o'clock.'

'Good; you'll be staying there again, then?'

'Well, I guess probably not,' he replied musingly. 'There were one or two things that, frankly, put me off. First of all, I couldn't lock my door, so I had to shut it and wedge a chair underneath the handle. Then I didn't feel too safe with that so, instead, I pushed a sideboard along inside the door. Then I found the air stifling, especially in the bedroom, and knowing that if I go to sleep with too much heat, I wake up with a bad headache, I tried to turn off the central heating. But I couldn't turn the handle and when I opened the window, it blew such an icy blast over my bed I thought that, if I left it like that, I would wake up with pneumonia at the least – and, in fact, I'd probably wake up dead! After all that exertion, and as I was feeling a little irritated, I thought I would calm down by having a bath. So I turned on the water in the bath, which had a mixer attachment, and found it came out boiling hot as the mixer wasn't working and there was no cold water. I ran the water in the hand basin to have a good wash and got both cold and hot water there – but the hot water tap got so hot that, when I came to turn it off, I burned my thumb! No, I guess they'll have to make one or two improvements before I feel like a night there again.'

Paul also received a large number of complimentary copies of books, sometimes from the authors (who were not necessarily even acquaintances of his, let alone friends) or from the publishers, hoping, no doubt, for some repeatable commendation. He always read the autobiographies written by those he knew and was usually enthralled by them. He loved to read about other people's success stories and how they overcame their own adversities, built-up their respective business empires and, in particular, how they dealt with any legal suits or difficulties with government agencies and departments and which ministries were helpful in solving what crises. These facts he kept in his head and he would often produce them years later, apparently out of the blue, when confronted in his own business by a parallel set of circumstances.

He rarely had the time or the inclination to read the other books sent to him and they were usually divided between the main library and the bookcases in the private office, which consequently ended up with an extraordinarily wide selection of literature, including copies of the various books by or about Paul which he, in return, would occasionally present to some visitor.

Other gifts which arrived were always, if possible, returned. There was a large amount of self-protection in this, as fairly early on, he was caught by someone sending him a 'gift' and, a few months later, writing peevishly to ask when it was going to be returned, saying that it had only been 'lent' for inspection. (This was an old trick – if the object in question was not returned, the owner would ask for financial compensation and threaten legal action!) Luckily, this gift had been packed away safely in the store-room. In the case of presents such as teddy bears, other soft toys, pencils and so on, these were handed on to a local children's hospital.

Where, for one reason or another, they were simply not returnable, gifts like gloves, socks and scarves were quietly handed on – without any publicity – to the local old people's homes or clubs.

Paul grew quite apprehensive about any mention in the newspapers of what he liked to have as presents or what he was expecting, or had had, for his birthday or Christmas. If he said a pair of socks, then dozens would arrive through the post from all over the world – short, long, patterned, plain, home-knitted or chain-store bought. One year it was pencils by the score, another, gloves and scarves.

For one of his birthdays, I gave him a little velvet-covered toy lion which he kept among his other bric-à-brac in the bathroom. When *Paris-Match* came to take photographs for an article about him and Sutton Place, the photographer found the lion and placed it on the corner of the bath so that it was right in the foreground of the view of the bathroom. Within two weeks of the publication of the article, toy lions (all sizes) began to arrive. In most cases, it was just not possible to return them to the senders – but the children in the local hospital must have thought they were in the Lion's House at the zoo by the time we had sent on several dozen lions from almost as many countries.

'If you give me anything else,' said Paul, 'I wish

that you'd make it something people can't send me more of.'

After that, I always tried to give him something that was not easily copied and, contrary to so many reports, he was not a difficult person for whom to choose a present. No one, not even a billionaire, really has everything and it is merely a matter of selecting a gift which is of some particular interest to the individual. Paul had such wide interests that it was easy. He was thrilled with a beautiful Rosenthal china Alsatian which I found; the factory had destroyed the original mould so that it would never be repeated – and it is not so easy to find good models of Alsatians. Another time, I gave him a three-dimensional scale model of Sutton Place which was beautifully hand-made by the late Hal Broun-Morison of Finderlie (who used to make exquisite models for museums of such things as 'The Old Potter's Workshop' or 'Glass-blowing in the Middle Ages').

Unfortunately, both these presents were displayed at some risk to their safety, as one of Paul's particularly jealous women friends, who considered me, quite mistakenly, as a rival, threatened to smash them to pieces. Paul was obliged to warn her that if she laid so much as a finger on either, she would be out of the house for good.

Then my mother suggested that, as he hated to sit for even a photographic portrait, and never did at all for a painting, he might like to have a sketch from me. I knew him and his features so well that I could draw him from memory, filling in any missing details the next time I saw him. He thought the first pencil sketch made him look older than his age and rather like a bulldog but he came to like it almost the best in the

173

end, saying that he had 'grown into it'! I did several more of him in the following years and, as far as I know, they were the only portraits from life of him during the last fifteen years of his life.

One particular birthday of his, we met for lunch and I gave him his latest portrait, a watercolour profile. He seemed very pleased and kept on looking at it.

We later went to Wildenstein's Gallery in Bond Street where they had a very large old master painting on view for Paul's inspection. Paul walked in, still clutching my picture, re-wrapped, under his arm and shook hands with the various art experts who were standing around at attention.

We were placed in two chairs in front of a vast Reubens-type picture while all the experts enthused about the quality, the colours, the movement, the style and so on. Paul sat, staring at the picture but never uttering a word. Eventually, when the artistic stream, so to speak, had run out, Paul stood up and smiled very pleasantly, thanked everyone, saying he was most interested and grateful for having been allowed to see the picture. Hardly pausing for breath, he then bent down and picked up the parcel containing his portrait, drew the painting out with great pride, and to my considerable embarrassment, said, 'How do you gentlemen all like my new picture that Miss Lund has just painted of me? I think it's a very good likeness, don't you?'

After we had left, I said to him, 'Oh, Paul! Fancy showing my small picture to all those experts when you were supposed to be viewing an Old Master worth well over half a million pounds!'

'I don't see why not,' he replied. 'Anyway, they may be able to criticise and dissect other people's paintings but I'll bet you none of them can paint a recognisable picture of anything themselves.'

'That reminds me,' I said, 'you never spoke a word while they were extolling the qualities of that picture. You sat contemplating it for ages but never said whether you really liked it or if you agreed with their views. What were you thinking all that time?'

'If you want to know, I was working out from the size of the canvas and the price that's being asked, how much it would cost per square inch and I came to the conclusion it was too much!'

Occasionally, Paul indulged in a spontaneous gesture of generosity: usually everything he did was pre-meditated and on-the-spur-of-the-moment decisions were not his style.

One year, he and I went to the private view of the Royal Academy of Art's Summer Exhibition at Burlington House. As we went in, we met my mother, who had just finished going round the galleries.

'Did you see anything you liked, Catherine dear?' asked Paul.

'Not much,' replied my mother, 'except for one painting in the South room which, although it is modern, I did like. It is intriguing and very clever.'

'Oh,' Paul asked with interest, 'which one is that?'

'You won't mistake it,' she said. 'There's a crowd round it, trying to touch it, and it's by someone called Jones. I won't tell you any more – you can go and see it for yourselves.'

So, having collected our catalogues, I looked up the South room and found the picture listed. There certainly was a crowd bunched together in one corner of the rather small room. As we worked our way closer, we got our first view and, sure enough, two or three people were touching what appeared to be a collage consisting of a paperback book, a gold pocket watch and a piece of twisted red cord hanging from a hook, all fixed to a clip-board.

'Well!' I said in some astonishment to Paul, 'my mother doesn't usually like collages of that type – she considers they're not art at all.'

Puzzled and intrigued, we moved nearer and, like so many others, reached forward to smooth out the turned back corner of the paperback – a copy of *The Nude* by Kenneth Clark. Only – there was no book, no watch, no cord, no hooks, even; the whole apparent montage was so cleverly painted with a three-dimensional effect that it was the perfect trompe l'oeil! Everyone who looked at it was taken in by it.

Paul was most impressed and spent the rest of our tour around the exhibition remarking that he did not care for this or that picture (frequently by well-known artists) nearly as well as that other one by Mr Jones. When we were about to leave, he took my arm and guided me over to the sales counter.

'Come on,' he said, 'I'm going to buy that picture.'

He went through all the formalities and, as we left, I asked him, 'You really liked it then?'

'I certainly do. I think it is very well executed technically and I like the joke it plays on its viewers.'

'But where are you going to hang it? It's rather different from your Bonnards, Renoirs and Rembrandts!'

'Oh, I've bought it for your mother,' he said. 'She said she liked it and I agree, so I thought it would be a nice little surprise for her.'

It certainly was – and the picture is on one of our walls where it is seen by all our guests, so many of whom have reached out to flatten that curling corner of the book.

'If I had nothing to do in the way of work and no responsibilities except to myself,' Paul said several times, 'I would like to be a beachcomber – or, to be quite perfect, a beachcomber with one luxury, a library! I would be quite happy to spend all day on the beach, eating avocados and reading. I would have all my Henty books for exciting reading, a few well-written biographies and some books to amuse me. I would choose biographies about interesting people, hoping that the author was not more concerned, as is usually the case, with showing off the background research he has done or his own clever assessment of his subject's character, than with facts. Amusing or witty books are hard to find these days. Modern authors seem incapable of competing with the shrewd erudition and pungent observations of, say, Oscar Wilde or Henry James.'

Paul had become an enthusiastic reader of Henry James ever since my mother and I took him to see *The Aspern Papers* and he had sat enthralled throughout the performance – his taste in theatre normally being towards the lighter side of the revue or the musical comedy type.

Latterly, he used to enjoy hearing odd pieces of the satirical verse which my mother wrote about business, social life, politics and so on.

'I'm getting to be so old,' he would say, 'that they'll make me retire soon, and then I shall be able to go beachcombing – with my books along with me, of course! You'd better tell your mother, dear, to hurry up and get her verse published as a book, then I can take it with me.'

177

One day, I was able to tell him that she was going to have it published in about six months' time.

'Fine,' he said, 'put me on the list of subscribers.'

In fact he never saw the book, although he had enquired after its progress so often and with such interest. He died two weeks after its publication and before I could get a copy to him.

If Paul had a film star pin-up, it was Greta Garbo. Whenever one of her films was being shown, he would make a special point of coming up to town to see it, sometimes even cancelling other appointments. When, some years ago, there was a complete 'Garbo' season in London, he telephoned me with a list of dates and times when the various films were being shown and arranged about fourteen visits for us to go to the cinema, as some films he wanted to see twice or more. *Queen Christina*, which was my favourite as well as his, we saw four times.

Once, when we were driving across France, he discovered (and I never found out how) that the cinema in a small town about thirty miles off our route to the north, was showing *Camille*, so we made a detour, left the car outside the police station (there was nowhere else to park), and spent the afternoon in a rather uncomfortable, noisy, garlicy cinema – and thoroughly enjoyed ourselves!

In London, his favourite restaurant for lunch was Trader Vic's at the Hilton Hotel. I am not sure if he went there for the food or the rum-based drinks to which he was very partial, his choice usually being a Mai Tai or a Planter's Punch.

In the evening, he liked the Grill Room of the Connaught Hotel – in fact he liked the whole atmosphere of the hotel and always stopped to admire the carving

178

and polish of the magnificent main staircase.

For dinner and dancing – of which he was very fond – he liked the Terrace Room of the Dorchester Hotel and his favourite night-club, while it existed, was the '400' in Leicester Square. He loved ballroom dancing so much that, with an orchestra or band that he liked, he would sometimes stay on the dance floor for an hour or more without stopping. It was frequently I who said, 'Shall we go back to the table and have a short rest?'

Normally, I was given about ten minutes sitting down and then he would hear some dance tune he liked – such as *Body and Soul*, which was one of his favourites – and immediately he would stand up again and say, 'Come on, Robina, we'll dance this one,' and off we would go for the next half hour or so.

He liked to learn new steps and I taught him in succession some simple jiving, the twist and the cha-cha. He, in his turn, devised some quite complicated rumba and samba routines which occasionally led to his disappearing across the dance floor while I was somehow supposed to encircle it, (while still dancing the steps), and meet him the other side. The result was that I usually lost him and would wait at the side of the dance floor, anxiously looking around, only to be tapped smartly on the shoulder by him after he had taken a great deal of trouble to creep up unseen behind me!

Paul, who had a great admiration for anyone who actually *did* anything as opposed to being a critic or expert about other people's abilities, was very inquisitive and liked to know what one was doing. He insisted that I showed him the pictures that I drew or painted, which were usually portraits of humans or animals and, if he liked one particularly, I usually gave it to him.

In the autumn of 1974, when the vogue for knitted and crocheted hats came in, I started my own designing and making-up in order to raise money for charity. Some hats I sold through Norman Hartnell Ltd. – from whom I had bought several most beautiful evening dresses – and other I sold direct to customers, all the money for them being paid direct to the charity concerned. Surprisingly quickly, orders seemed to mount up and it was just as well that I do not sleep much, as the only time I had to make them was between about eleven at night and two or three in the morning. Paul, who always worked late himself, though he did not get up at a quarter past seven like me, used quite often to ring me at midnight or one in the morning to see how I was getting on. He had for years telephoned at odd hours, often very late at night, in order to have some half-hour or hour-long business discussion or to mull over his latest art acquisition. Now, however, he was as keen as I was to get the hat orders completed and would consider what I had finished and what was left to be done.

Then he would remark, 'We'll leave the business until tomorrow, dear, and that means you can do two or three more hats by half past two. That's the Italian order complete then, isn't it?'

By now, I was making in (for me) bulk and exporting to the United States, Canada, France, Switzerland, Italy and Norway and, when winter was over, I changed materials and started on summer hats. Paul tried to persuade me to take on knitters and outside workers until I explained that everything was in my head. I didn't know how technically to write down the patterns I was making and, anyway, they were not very easy to copy and having nothing in writing meant that there was more chance of the designs remaining exclusive.

'In that case,' said Paul, 'you had better teach me what to do, then I can always fill in for you.'

I did not really expect him ever to start making crocheted hats but he did learn how to wield a crochet pin and was very proud of his ability.

When I went down to Sutton, I would take a box of assorted sample hats with me (large sizes) which we would unpack in his study. Paul would then fool around trying them on, brims up, brims down, back-to-front, or pulled well down to his eyebrow level! This would be accompanied by a mincing walk with undulating hips, (which could not have been more unlike his normal, slightly bent-kneed gait), my handbag swinging on his arm and an attempt to flutter his short eyelashes.

He was always quite sorry to see everything packed up again and removed and, although I offered to make him a suitable hat ('with bobbles on?' he asked) for winter, he was never quite sure that he could break his many-years-old habit of never wearing a hat.

It was in 1966 that the *Daily Express* decided to run a competition in which one had to select, in order of importance out of a specified list, what one considered to be the essential attributes in order to become a self-made millionaire. The prize was £200 per week for a year.

Paul's eyes lit up. 'I wouldn't mind £200 a week for a year – in cash,' he remarked. 'Do you think it's tax free?'

He mulled over the list and then decided to go in

181

for the competition – under the assumed name which he occasionally used for his own convenience. At last, he produced his list triumphantly.

I read it through and could not resist asking him, 'Are these characteristics what you think the self-made millionaire ought to have – or how you would describe yourself?'

He chuckled. 'That list is certainly not me. I'll make another one of how I see myself but it certainly won't include many of the qualities listed by the *Express.*'

When he had finished, I compared the two lists. The 'ideal' characteristics he had chosen to be a self-made millionaire he had put down as:

Single-mindedness
Knowledge of particular business
Ability to make good use of professional advice
Knowledge of finance
Attention to detail
Forward-looking
Cool nerve in crisis
Receptive of new ideas and methods

His own attributes he had assessed as:

Single-mindedness
Ability to work long hours (18+ hours a day, 7 days a week if necessary)
Attention to detail
Knowledge of particular business
Unpredictability (stops business rivals anticipating decisions and staff from getting slack)
Elastic use of time (time should be made to adapt to the work to be done, not work to the time allotted;

no one can work at his best if he works at a continuously full gallop)
Ability to make good use of professional advice

We sent in his 'official' entry, with slight variations
in the order of importance of his selected qualities and
awaited, with great interest, the result of the competition.

As the day of the announcement came, having
heard nothing, we guessed that he was not to be the
lucky winner but were intrigued to see how the chosen
list, selected, incidentally, by a panel of five judges,
varied from Paul's. When we read through it, we both
had a few surprises and could hardly stop laughing
because some of it was almost the perfect antithesis
of Paul. It went as follows:

Single-mindedness
Forceful and ambitious personality
Forward looking
Cool nerve in crisis
Quick and firm in decisions
Sound judgement of staff
Knowledge of finance
Full, economical use of time

Paul, his remarks accompanied by various grunts
and humphs, proffered the following comments, 'Ambitious personality is all right, but being forceful
doesn't necessarily get you anywhere but backwards if
you put the nose of someone who could help you out
of joint. I prefer the "softly, softly" method.

'Cool nerve in crisis? Well, most crises have to be
solved by the man on the spot – he's the one that needs
the cool nerve.

'Quick and firm in decisions – no! I think the better policy generally is to decide slowly, move quickly, and be flexible rather than firm. Professional advisers are, I reckon, more important than staff.

'And, finally, I've never used my time, as I said before, fully or economically, by intention, except when the work demanded it. I don't believe in nine-to-five hours. If you have four hours of work to do, why sit in an office for eight hours? On the other hand, if you have fifteen hours of work to get done, you start before nine in the morning and you don't close up shop at five o'clock.'

He glanced back at the competition result and started to chuckle. 'I think I'll write to the Editor of the *Daily Express* and say I'm really disappointed to learn that I'm just never going to succeed as a self-made millionaire!'

POOR LITTLE RICH MAN

Poor little rich man, how sad is his life,
He's tops with his girl-friends but not with his wife.
He buys up the market, both Common and Stock,
And takes over companies by buying en bloc.
His assets abroad, in fashionable places,
Provide what he needs in luxury bases.
When not in New York, or Paris, or Rome,
Perish the thought, but he might be at home;
This is but a flash in his gold-mining pan,
Then he gets back to work as fast as he can.

   * * *

Poor little rich man, how sad is his life.